Principles of Data Fabric

Become a data-driven organization by implementing Data Fabric solutions efficiently

Sonia Mezzetta

BIRMINGHAM—MUMBAI

Principles of Data Fabric

Publishing Product Manager: Heramb Bhavsar

Content Development Editor: Manikandan Kurup

Technical Editor: Kavyashree K S

Copy Editor: Safis Editing

Project Coordinator: Farheen Fathima

Proofreader: Safis Editing

Indexer: Pratik Shirodkar

Production Designer: Prashant Ghare

Marketing Coordinators: Nivedita Singh

First published: April 2023

Production reference: 1310323

Published by Packt Publishing Ltd.

Livery Place

35 Livery Street

Birmingham

B3 2PB, UK.

ISBN 978-1-80461-522-5

www.packtpub.com

To my daughter, Melania; you are and will always be my forever inspiration in everything I do. You are my hero. I know it wasn't easy at times not having my undivided attention, so thank you for your patience while I wrote this book.

To Mike, thank you for your love and significant support throughout this journey. I couldn't have done this without your help.

To my parents and sisters, thank you for always being there for me and for rooting me on. Family is everything.

To my loving pets at home, Cody and Stella, and to those pets no longer with us, Bella and Lobo. I miss you both dearly.

– Sonia Mezzetta

Contributors

About the author

Sonia Mezzetta is a senior certified IBM architect working as a Data Fabric program director. She has an eye for detail and enjoys problem-solving data pain points. She started her data management career at IBM as a data architect. She is an expert in Data Fabric, DataOps, Data Governance, and Data Analytics. With over 20 years of experience, she has designed and architected several enterprise data solutions. She has authored numerous data management white papers and has a master's and bachelor's degree in computer science. Sonia is originally from New York City and currently resides in the area of Westchester County, New York.

About the reviewers

Lena Woolf is an experienced senior technical staff member of IBM Watson Knowledge Catalog, Data, and AI. Over the course of her IBM career, Lena has made sustained technical contributions to IBM software products and has been recognized as a deep technical expert and thought leader in the fields of information management and governance. Lena regularly speaks at professional conferences. As an inventor, she has contributed to many patents and constantly pushes the boundaries of what's possible with IBM technology. Lena is passionate about providing a gender-inclusive workspace where everyone can collaborate to drive innovation and business growth.

Jo Ramos is a distinguished engineer and the Director and Chief Solutions Architect for Data Fabric at IBM Expert Labs. Jo leads the technology and platform architecture team to support clients on their data modernization and transformation journey to accelerate the adoption of data and AI enterprise capabilities. Jo has extensive experience working as a technologist and thought leader across multiple industries, designing innovative data and analytics solutions for enterprises. His specialties includes Data Fabric, Data Mesh, DataOps, Data Governance, Data Integration, Big Data, Data Science and Analytics.

Rosalind Radcliffe is an IBM Fellow and CIO DevSecOps CTO. Rosalind is responsible for driving DevSecOps and application modernization transformation for the IBM CIO office with the goal of making the office the showcase for hybrid cloud. In this role, she works with the CIO office and partners on research and development to drive the adoption of common practices and tools. Ultimately, this effort will transform, standardize, and automate the processes, tools, and methodologies used to make IBM the most secure, agile, efficient, and automated hybrid cloud engineering organization. In her prior role, she was responsible for bringing open modern toolchains to the z/OS platform and working with clients on their DevOps transformation. She is a frequent speaker at conferences, a master inventor, a member of the IBM Academy of Technology, and the author of *Enterprise Bug Busting*.

Table of Contents

Part 1: The Building Blocks

1

2

Part 2: Complementary Data Management Approaches and Strategies

3

Choosing between Data Fabric and Data Mesh   31

4

Introducing DataOps   45

Preface

Data constitutes facts, statistics, and information based on real-world entities and events. The word *fabric* represents a body of material with texture and structure, such as silk cloth. These two keywords, *Data Fabric*, create a representation of disparate data that has been connected by a data architecture driven by governance, active metadata, automated data integration, and self-service. In today's big data era, there are many complexities faced by enterprises looking to become data driven. Many of these issues, such as data silos, agility, lack of collaboration between business and IT, high maintenance costs, data breach, and data integrity, revolve around the large volume and velocity of proliferated data. Data Fabric is a mature, composable data architecture that faces these complexities head-on to enable the management of data at a high scale with established business value.

I wrote this book to introduce a slightly different perspective on the definition of Data Fabric architecture. The view I offer is flexible and use case agnostic and supports diverse data management styles, operational models, and technologies. I describe Data Fabric architecture as taking a people, process, and technology approach that can be applied in a complementary manner with other trending data management frameworks, such as Data Mesh and DataOps. The main theme of this book is to provide a guide to the design of Data Fabric architecture, explain the foundational role of Data Governance, and provide an understanding of how Data Fabric architecture achieves automated Data Integration and Self-Service. The technique I use is by describing "a day in the life of data" as it steps through the phases of its life cycle: create, ingest, integrate, consume, archive, and destroy. I talk about how each layer in Data Fabric architecture executes in a high-performing and thorough manner to address today's big data complexities. I provide a set of guidelines, architecture principles, best practices, and key concepts to enable the design and implementation of a successful Data Fabric architecture.

The perspective I offer is based on decades of experience in the areas of Enterprise Architecture, Data Architecture, Data Governance, and Product Management. I remember when I started my career in Data Governance, I faced many challenges convincing others of the business value that successful data management with Data Governance achieves. I saw what many others failed to see at that time, and that was when I knew data was my passion! Since then, I've broadened and increased my knowledge and experience. I have learned from brilliant thought leaders at IBM and a diverse set of clients. All these experiences have shaped the frame of reference in this book.

As technologists, we are very passionate about our points of view, ideas, and perspectives. This is my point of view on what a Data Fabric architecture design represents, which aims to achieve significant business value while addressing the complexities enterprises face today.

> **Note**
> The views expressed in the book belong to the author and do not necessarily represent the opinions or views of their employer, IBM.

Who this book is for

This book is for an organization looking to venture on a digital transformation journey, or an existing data-driven organization looking to mature further in their data journey. It is intended for a diverse set of roles, both business and technical, with a vested interest in strategic, automated, and modern data management, including the following:

- Executive leaders such as chief data officers, chief technology officers, chief information officers, and data leaders prioritizing strategic investments to execute an enterprise data strategy

- Enterprise architects, data architects, Data Governance roles such as data security, data privacy roles, and technical leaders tasked with designing and implementing a mature and governed Self-Service data platform

- Business analysts and data scientists looking to understand their role as data producers or data consumers in a Self-Service ecosystem leveraging Data Fabric architecture

- Developers such as data engineers, software engineers, and business intelligence developers looking to comprehend Data Fabric architecture to learn how it achieves the rapid development of governed, trusted data

What this book covers

Chapter 1, Introducing Data Fabric, presents an introduction to the definition of Data Fabric architecture. It offers a position on what Data Fabric is and what it isn't. Key characteristics and architectural principles are explained. Essential concepts and terminology are defined. The business value statement of Data Fabric architecture is discussed and the core building blocks that make up its design are established.

Chapter 2, Show Me the Business Value, is a chapter focused on providing a business point of view on the benefits of Data Fabric architecture. It establishes the business value the architecture offers by explaining how the building blocks that make up a Data Fabric design address pain points faced by enterprises today. Data Fabric architecture takes a strategic and multi-faceted approach to achieve data monetization. Real-life examples have been positioned on the impact of not having the right level of focus provided by each of Data Fabric's building blocks. Finally, a perspective is offered on how Data Fabric architecture can be leveraged by large, medium-sized, and small organizations.

Chapter 3, Choosing between Data Fabric and Data Mesh, provides an overview of the key principles of Data Mesh architecture. Both Data Fabric and Data Mesh are discussed, including where they share similar objectives and where they take different but complementary approaches. Both architectures represent sophisticated designs focused on data trust and enable the high-scale sharing of quality data.

This chapter closes with a view on how Data Fabric and Data Mesh can be used together to achieve rapid data access, high-quality data, and automated Data Governance.

Chapter 4, Introducing DataOps, introduces the DataOps framework. It discusses the business value it provides and describes the 18 driving principles that make up DataOps. The role of data observability and its relationship to the Data Quality and Data Governance pillar is explained. This chapter concludes by explaining how to apply DataOps as an operational model for Data Fabric architecture.

Chapter 5, Building a Data Strategy, kicks off the creation and implementation of a data strategy document. It describes a data strategy document as a visionary statement and a plan for profitable revenue and cost savings. You will familiarize yourself with the different sections that should be defined in a data strategy document, and have a reference of three data maturity frameworks to use as input in a data strategy. The chapter ends with tips on how Data Fabric architecture can be positioned as part of a data strategy document.

Chapter 6, Designing a Data Fabric Architecture, sets the foundation for the design of a Data Fabric architecture. It introduces key architecture concepts and architecture principles that compose the logical data architecture of a Data Fabric. The three architecture layers, Data Governance, Data Integration, and Self-Service, in a Data Fabric architecture are introduced. The objectives of each layer are highlighted, with a discussion on the necessary capabilities represented as components.

Chapter 7, Designing Data Governance, dives into the design of the Data Governance layer of a Data Fabric architecture. Key architecture patterns, such as metadata-driven and event-driven architectures, are discussed. The architecture components, such as active metadata, metadata knowledge graphs, and life cycle governance, are explained. The chapter ends with an explanation of how the Data Governance layer executes and governs data at each phase in its life cycle.

Chapter 8, Designing Data Integration and Self-Service, drills into the design of the two remaining architecture layers in a Data Fabric, Data Integration and Self-Service. The Data Integration layer is reviewed, which focuses on the development of data with a DataOps lens. The Self-Service layer is also discussed, including how it aims to democratize data. An understanding is provided of how both architecture layers work with each other, and how they rely on the Data Governance layer. At the end of the chapter, a Data Fabric reference architecture is presented.

Chapter 9, Realizing a Data Fabric Technical Architecture, positions a technical Data Fabric architecture as modular and composable, consisting of several tools and technologies. The required capabilities and the *kinds* of tools to implement each of the three layers in a Data Fabric architecture are discussed. Two use cases are reviewed – distributed data management via Data Mesh and regulatory compliance – as examples of how to apply a Data Fabric architecture. The chapter ends by presenting a Data Fabric with Data Mesh technical reference architecture.

Chapter 10, Industry Best Practices, presents 16 best practices in data management. Best practices are grouped into four categories: Data Strategy, Data Architecture, Data Integration and Self-Service, and Data Governance. Each best practice is described and has a *why should you care* statement.

To get the most out of this book

To understand the key concepts and themes in this book, you should have a general understanding of the IT industry, enterprise architectures, data management, and Data Governance.

Conventions used

There are a number of text conventions used throughout this book.

Bold: Indicates a new term, an important word, or words that you see onscreen. For instance, words in menus or dialog boxes appear in **bold**. Here is an example: "Select **System info** from the **Administration** panel."

> **Tips or important notes**
> Appear like this.

Get in touch

Feedback from our readers is always welcome.

General feedback: If you have questions about any aspect of this book, email us at `customercare@packtpub.com` and mention the book title in the subject of your message.

Errata: Although we have taken every care to ensure the accuracy of our content, mistakes do happen. If you have found a mistake in this book, we would be grateful if you would report this to us. Please visit `www.packtpub.com/support/errata` and fill in the form.

Piracy: If you come across any illegal copies of our works in any form on the internet, we would be grateful if you would provide us with the location address or website name. Please contact us at `copyright@packt.com` with a link to the material.

If you are interested in becoming an author: If there is a topic that you have expertise in and you are interested in either writing or contributing to a book, please visit `authors.packtpub.com`.

Share Your Thoughts

Once you've read *Principles of Data Fabric*, we'd love to hear your thoughts! Scan the QR code below to go straight to the Amazon review page for this book and share your feedback.

https://packt.link/r/1804615226

Your review is important to us and the tech community and will help us make sure we're delivering excellent quality content.

Download a free PDF copy of this book

Thanks for purchasing this book!

Do you like to read on the go but are unable to carry your print books everywhere?

Is your eBook purchase not compatible with the device of your choice?

Don't worry, now with every Packt book you get a DRM-free PDF version of that book at no cost.

Read anywhere, any place, on any device. Search, copy, and paste code from your favorite technical books directly into your application.

The perks don't stop there, you can get exclusive access to discounts, newsletters, and great free content in your inbox daily

Follow these simple steps to get the benefits:

1. Scan the QR code or visit the link below

https://packt.link/free-ebook/9781804615225

2. Submit your proof of purchase

3. That's it! We'll send your free PDF and other benefits to your email directly

Part 1:
The Building Blocks

Data Fabric architecture, alongside its distinguishing qualities and business value proposition, needs to first be defined to enable its adoption as part of any data management strategy.

The first part of this book introduces Data Fabric architecture by establishing the core building blocks and their business value proposition. I offer a different perspective on what defines Data Fabric architecture than ones on the market today. Data Fabric is a flexible and composable architecture capable of adopting several data management styles and operational models. Foundational Data Governance pillars and their intended focus are explained, and a list of key characteristics of what does and doesn't define Data Fabric is provided.

By the end of *Part 1*, you will have an understanding of what Data Fabric architecture is, its differentiating characteristics and architecture principles, and the impact of not having a Data Governance-centric architecture.

This part comprises the following chapters:

- *Chapter 1, Introducing Data Fabric*
- *Chapter 2, Show Me the Business Value*

1

Introducing Data Fabric

Data Fabric is a distributed data architecture that connects scattered data across tools and systems with the objective of providing governed access to fit-for-purpose data at speed. Data Fabric focuses on Data Governance, Data Integration, and Self-Service data sharing. It leverages a sophisticated active metadata layer that captures knowledge derived from data and its operations, data relationships, and business context. Data Fabric continuously analyzes data management activities to recommend value-driven improvements. Data Fabric works with both centralized and decentralized data systems and supports diverse operational models. This book focuses on Data Fabric and describes its data management approach, differentiating design, and emphasis on automated Data Governance.

In this chapter, we'll focus on understanding the definition of Data Fabric and why it's important, as well as introducing its building blocks. By the end of this chapter, you'll have an understanding of what a Data Fabric design is and why it's essential.

In this chapter, we'll cover the following topics:

- What is Data Fabric?
- Why is Data Fabric important?
- Data Fabric building blocks
- Operational Data Governance models

> **Note**
> The views expressed in the book belong to the author and do not necessarily represent the opinions or views of their employer, IBM.

What is Data Fabric?

Data Fabric is a distributed and composable architecture that is metadata and event driven. It's use case agnostic and excels in managing and governing distributed data. It integrates dispersed data with automation, strong Data Governance, protection, and security. Data Fabric focuses on the Self-Service delivery of governed data.

Data Fabric does not require the migration of data into a centralized data storage layer, nor to a specific data format or database type. It can support a diverse set of data management styles and use cases across industries, such as a 360-degree view of a customer, regulatory compliance, cloud migration, data democratization, and data analytics.

In the next section, we'll touch on the characteristics of Data Fabric.

What Data Fabric is

Data Fabric is a composable architecture made up of different tools, technologies, and systems. It has an active metadata and event-driven design that automates Data Integration while achieving interoperability. Data Governance, Data Privacy, Data Protection, and Data Security are paramount to its design and to enable Self-Service data sharing. The following figure summarizes the different characteristics that constitute a Data Fabric design.

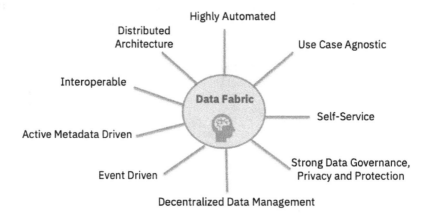

Figure 1.1 – Data Fabric characteristics

Data Fabric takes a proactive and intelligent approach to data management. It monitors and evaluates data operations to learn and suggest future improvements leading to productivity and prosperous decision-making. It approaches data management with flexibility, scalability, automation, and governance in mind and supports multiple data management styles. What distinguishes Data Fabric architecture from others is its inherent nature of embedding Data Governance into the data life cycle as part of its design by leveraging metadata as the foundation. Data Fabric focuses on business controls with an emphasis on robust and efficient data interoperability.

In the next section, we will clarify what is *not* representative of a Data Fabric design.

What Data Fabric is not

Let's understand what Data Fabric is not:

- It is not a single technology, such as data virtualization. While data virtualization is a key Data Integration technology in Data Fabric, the architecture supports several more technologies, such as data replication, ETL/ELT, and streaming.

- It is not a single tool like a data catalog and it doesn't have to be a single data storage system like a data warehouse. It represents a diverse set of tools, technologies, and storage systems that work together in a connected ecosystem via a distributed data architecture, with active metadata as the glue.

- It doesn't just support centralized data management but also federated and decentralized data management. It excels in connecting distributed data.

- Data Fabric is not the same as Data Mesh. They are different data architectures that tackle the complexities of distributed data management using different but complementary approaches. We will cover this topic in more depth in *Chapter 3, Choosing between Data Fabric and Data Mesh*.

The following diagram summarizes what Data Fabric architecture does not constitute:

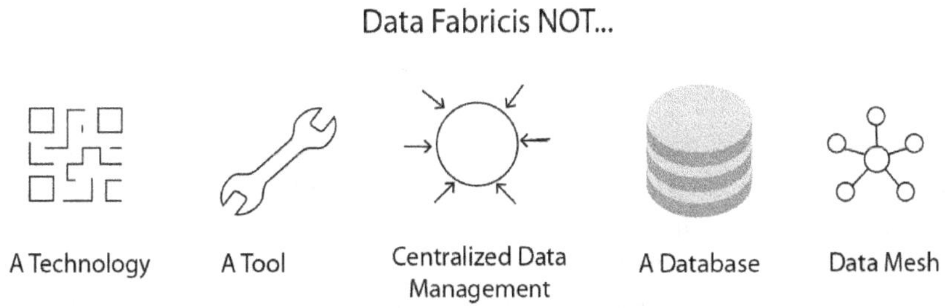

Data Fabricis NOT...

| A Technology | A Tool | Centralized Data Management | A Database | Data Mesh |

Figure 1.2 – What Data Fabric is not

We have discussed in detail what defines Data Fabric and what does not. In the next section, we will discuss why Data Fabric is important.

Why is Data Fabric important?

Data Fabric enables businesses to leverage the power of connected, trusted, protected, and secure data no matter where it's geographically located or stored (cloud, multi-cloud, hybrid cloud, on-premises, or the edge). Data Fabric handles the diversity of data, use cases, and technologies to create a holistic end-to-end picture of data with actionable insights. It addresses the shortcomings of previous data management solutions while considering lessons learned and building on industry best practices. Data Fabric's approach is based on a common denominator, metadata. Metadata is the secret sauce of Data Fabric architecture, along with automation enabled by machine learning and **artificial intelligence**

(**AI**), deep Data Governance, and knowledge management. All these aspects lead to the efficient and effective management of data to achieve business outcomes, therefore cutting down on operational costs and increasing profit margins through strategic decision-making.

Some of the key benefits of Data Fabric are as follows:

- It addresses data silos with actionable insights from a connected view of disparate data across environments (cloud, multi-cloud, hybrid cloud, on-premises, or the edge) and geographies

- Data democratization leads to a shorter time to business value with frictionless Self-Service data access

- It establishes trusted, secure, and reliable data via automated Data Governance and knowledge management

- It enables a business user with intuitive discovery, understanding, and access to data while addressing a technical user's needs, supporting various data processing techniques in order to manage data. Such approaches are batch or real time, including ETL/ELT, data virtualization, change data capture, and streaming

Now that we have a view of why Data Fabric is important and how it takes a modern approach to data management, let's review some of the drawbacks of earlier data management approaches.

Drawbacks of centralized data management

Data is spread everywhere: on-premises, across cloud environments, and on different types of databases, such as SQL, NoSQL, data lakes, data warehouses, and data lakehouses. Many of the challenges associated with this in the past decade, such as data silos, still exist today. The traditional data management approach to analytics is to move data into a centralized data storage system. Moving data into one central system facilitates control and decreases the necessary checkpoints across the large number of different environments and data systems. Thinking about this logically, it makes total sense. If you think about everyday life, we are successful at controlling and containing things if they are in one central place.

As an example, consider the shipment of goods from a warehouse to a store that requires inspection during delivery. Inspecting the shipment of goods in one store will require a smaller number of people and resources as opposed to accomplishing this for 100 stores located across different locations. Seamless management and quality control become a lot harder to achieve across the board. The same applies to data management, and this is what led to the solution of centralized data management.

While centralized data management was the de facto approach for decades and is still used today, it has several shortcomings. Data movement and integration come at an expensive cost, especially when dealing with on-premises data storage solutions. It heavily relies on data duplication to satisfy a diverse set of use cases requiring different contexts. Complex and performance-intensive data pipelines built to enable data movement require intricate maintenance and significant infrastructure investments, especially if automation or governance is nowhere in the picture. In a traditional operating model, IT departments centrally manage technical platforms for business domains. In the past and still today, this model creates bottlenecks in the delivery of and access to data, minimizing the time to value.

Enterprise data warehouses

Enterprise data warehouses are complex systems that require consensus across business domains on common definitions of data. An enterprise data model is tightly coupled to data assets. Any changes to the physical data model without proper dependency management breaks downstream consumption. There are also challenges in Data Quality, such as data duplication and the lack of business skills to manage data within the technical platform team.

Data lakes

Data lakes came after data warehouses to offer a flexible way of loading data quickly without the restrictions of upfront data modeling. Data lakes can load raw data as is and later worry about its transformation and proper data modeling. Data lakes are typically managed in NoSQL databases or file-based distributed storage such as Hadoop. Data lakes support semi-structured and unstructured data in addition to structured data. Challenges with data lakes come from the very fact that they bypass the need to model data upfront, therefore creating unusable data without any proper business context. Such data lakes have been referred to as *data swamps*, where the data stored has no business value.

Data lakehouses

Data lakehouses is a new technology and is a combination of both Data Warehouse and Data Lake design. Data lakehouses support structured, unstructured and semi-structured data and are capable of addressing data science and business intelligence use cases.

Decentralized data management

While there are several great capabilities in centralized data systems, such as data warehouses, data lakes, and data lakehouses, the reality is, we are at a time where *all* these systems have a role and create the need for decentralized data management. A single centralized data management system is not equipped to handle all possible use cases in an organization and at the same time excel in proper data management. I'm not saying there is no need for a centralized data system, but rather, it can represent a progression. For example, a small company might start with one centralized system that fits their business needs, and as they grow, they evolve into more decentralized data management.

Another example is a business domain within a large company that might own and manage a data lake, or a data lakehouse that needs to co-exist with several other data systems owned by other business domains. This again represents decentralized data management. Cloud technologies have further provoked the proliferation of data. There is a multitude of cloud providers with their own set of capabilities and cost incentives, leading to organizations having multi-cloud and hybrid cloud environments.

We have evolved from a world of centralized data management as the best practice to a world in which decentralized data management is necessary. There is a seat at the table for all types of centralized systems. What's important is for these systems to have a data architecture that connects data in an intelligent and cohesive manner. This means a data architecture with the right level of control and rigor while balancing quick access to trusted data, which is where Data Fabric architecture plays a major role.

In the next section, let's briefly discuss considerations in building Data Fabric architecture.

Building Data Fabric architecture

Building Data Fabric architecture is not an easy undertaking. It's not a matter of building a simple 1-2-3 application or applying specific technologies. It requires collaboration, business alignment, and strategic thinking about the design of the data architecture; the careful evaluation and selection of different tools, data storage systems, and technologies; and thought into when to buy or build. Metadata is the common thread that ties data together in a Data Fabric design. Metadata must be embedded into every aspect of the life cycle of data from start to finish. Data Fabric actively manages metadata, which enables scalability and automation and creates a design that can handle the growing demands of businesses. It offers a future-proof design that can grow to add subsequent tools and technologies.

Now, with this in mind, let's introduce a bird's-eye view of a Data Fabric design by discussing its building blocks.

Data Fabric building blocks

Data Fabric's building blocks represent groupings of different components and characteristics. They are high-level blocks that describe a package of capabilities that address specific business needs. The building blocks are Data Governance and its knowledge layer, Data Integration, and Self-Service. *Figure 1.3* illustrates the key architecture building blocks in a Data Fabric design.

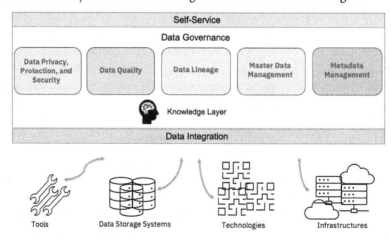

Figure 1.3 – Data Fabric building blocks

Data Fabric's building blocks have foundational principles that must be enforced in its design. Let's introduce what they are in the following subsection.

Data Fabric principles

Data Fabric's foundational principles ensure that the data architecture is on the right path to deliver high-value and high-quality data management that ensures data is secure, and protected. The following list introduces the principles that need to be incorporated as part of a Data Fabric design. In *Chapter 6, Designing a Data Fabric Architecture*, we'll discuss each principle in more depth:

- Data are Assets that can evolve into Data Products (TOGAF & Data Mesh): Represents a transition where assets have active product management across its life cycle from creation to end of life, specific value proposition and are enabled for high scale data sharing.

- Data is shared (TOGAF): Empower high-quality data sharing

- Data is accessible (TOGAF): Ease of access to data

- Data product owner (TOGAF and Data Mesh): The Data Product owner manages the life cycle of a Data Product, and is accountable for the quality, business value, and success of data

- Common vocabulary and data definitions (TOGAF): Business language and definitions associated with data

- Data security (TOGAF): Data needs to have the right level of Data Privacy, Data Protection and Data Security.

- Interoperable (TOGAF): Defined data standards that achieve data interoperability

- Always be architecting (Google): Continuously evolve a data architecture to keep up with business and technology changes

- Design for automation (Google): Automate repeatable tasks to accelerate time to value

These principles have been referenced directly or inspired the creation of new principles. from different noteworthy sources: TOGAF (https://pubs.opengroup.org/togaf-standard/adm-techniques/chap02.html#tag_02_06_02), Google (https://cloud.google.com/blog/products/application-development/5-principles-for-cloud-native-architecture-what-it-is-and-how-to-master-it), and Data Mesh, created by Zhamak Dehghani. They capture the essence of what is necessary for a modern Data Fabric architecture. I have slightly modified a couple of the principles to better align with today's data trends and needs.

Let's briefly discuss the four Vs in big data management, which are important dimensions that need to be considered in the design of Data Fabric.

The four Vs

In data management, the four Vs – Volume, Variety, Velocity, and Veracity (`https://www.forbes.com/sites/forbestechcouncil/2022/08/23/understanding-the-4-vs-of-big-data/?sh=2187093b5f0a`) – represent dimensions of data that need to be addressed as part of Data Fabric architecture. Different levels of focus are needed across each building block. Let's briefly introduce each dimension:

- **Volume**: The size of data impacts Data Integration and Self-Service approaches. It requires a special focus on performance and capacity. Social media and IoT data have led to the creation of enormous volumes of data in today's data era. The size of data is at an infinite point. Classifying data to enable its prioritization is necessary. Not all data requires the same level of Data Governance rigor and focus. For example, operational customer data requires high rigor when compared to an individual's social media status.

- **Variety**: Data has distinct data formats, such as structured, semi-structured, and unstructured data. Data variety dictates technical approaches that can be taken in its integration, governance, and sharing. Typically, structured data is a lot easier to manage compared to unstructured data.

- **Velocity**: The speed at which data is collected and processed, such as batch or real time, is a factor in how Data Governance can be applied and enabling Data Integration technologies. For example, real-time data will require a streaming tool. Data Governance aspects, such as Data Quality and Data Privacy, require a different approach when compared to batch processing due to its incomplete nature.

- **Veracity**: Data Governance centered around Data Quality and data provenance plays a major role in supporting this dimension. Veracity measures the degree to which data can be trusted and relied upon to make business decisions.

Now that we have a summary of the four Vs, let's review the building blocks (Data Governance, Data Integration, and Self-Service) of a Data Fabric design.

> **Important note**
> I have intentionally not focused on the tools or dived into the implementation details of Data Fabric architecture. My intention is to first introduce the concepts, groups of capabilities, and objectives of Data Fabric architecture at a bird's-eye-view level. In later chapters, we'll dive into the specific capabilities of each building block and present a Data Fabric reference architecture example.

Data Governance

Data Governance aims to define standards and policies that achieve data trust via protected, secure, and high-quality or fit-for-purpose data. Data Fabric enables efficient Data Integration by leveraging automation while making data interoperable. To support a mature Data Integration approach, the right level of Data Governance is required. It's of no use to have a design approach that beautifully integrates dispersed data only to find out that the cohesive data is of poor quality and massively violates data compliance stipulations. Costs saved in superior Data Integration would be short-lived if organizations are then slapped with millions of dollars in Data Privacy violation fines.

Data Governance isn't new; it has been around for decades. However, in the past, it was viewed as a burden by technologists and as a major obstacle in the deployment of successful software solutions. The impression has been that a governance authority sets rules and boundaries restricting progress with unnecessary rigor. What is drastically different today is the shift in perception. Data Governance has evolved to make it easy to access high-quality, trusted data via automation and more mature technologies despite the need to enforce security and regulation requirements. Another factor is the recognition of the impact of not having the right level of governance in a world where data managed in the right way can lead to major data monetization.

Data Governance today is recognized as a critical success factor in data initiatives. Data is predicted to grow at an exponentially fast rate. According to *IDC*, from 2021 to 2026, it will grow at a CAGR of 21.2%, potentially reaching 221,178 exabytes by 2026 (`https://www.idc.com/getdoc.jsp?containerId=US49018922`). This has pushed Data Governance to be front and center in achieving successful data management. To keep up with all the challenges introduced by big data, Data Governance has gone through a modernization evolution. In the following sections, we will dive into the pillars that represent Data Governance in Data Fabric architecture.

The Data Governance pillars established in the past are very relevant today with a modern twist. All Data Governance pillars are critical. However, Data Privacy, Protection and Security as well as Data Quality are prioritized due to data sovereignty requirements that need to address state-, country-, and union-based laws and regulations (such as the **General Data Protection Regulation** (**GDPR**) or the **California Consumer Privacy Act** (**CCPA**)), as well as the driving need for high-quality data to achieve data monetization. Data Lineage, Master Data Management, and Metadata Management are fundamental pillars that have dependencies on one another. The priority of these pillars will vary by use case and organizational objectives. All pillars serve a vital purpose and provide the necessary guardrails to manage, govern, and protect data.

Let's have a look at each Data Governance pillar.

Data Privacy, Protection, and Security

Whether data is used knowingly or unknowingly for unethical purposes, it needs to be protected by defining the necessary data policies and enforcement to automatically shield it from unauthorized access. Business controls need to be applied in an automated manner during Data Integration activities such as data ingestion, data movement, and data access. Manual or ad hoc enforcement is not feasible, especially when it comes to a large data volume. There need to be defined intelligent data policies that mask or encrypt data throughout its life cycle. There are several considerations that need to be taken regarding data access, such as the duration of access, data retention policies, security, and data privacy.

Data Quality

Data Quality was important 20 years ago and is even more critical today. This pillar establishes trust in using data to make business decisions. Today's world is measured by the speed of data monetization. How quickly can you access data – whether that's making predictions that generate sales via a machine learning model or identifying customer purchasing patterns via business intelligence reports? Data trust is an fundamental part of achieving data monetization – data needs to be accurate, timely, complete, unique, and consistent. A focus on Data Quality avoids data becoming stale and untrustworthy. Previous approaches only focused on Data Quality at rest, such as within a particular data repository that is more passive in nature. While this is still required today, it needs to be combined with applying Data Quality to in-flight data. Data Quality needs to keep up with the fluidity of data and take a proactive approach. This means Data Quality needs to be embedded into operational processes where it can be continuously monitored. This is called **data observability**.

Data Lineage

Data Lineage establishes confidence in the use of data. It accomplishes this by providing an audit trail starting with the data source, target destinations, and changes to data along the way, such as transformations. Data Lineage creates an understanding of the evolution of data throughout its life cycle. Data provenance is especially important for demonstrating compliance with regulatory policies. Data Lineage must keep track of the life cycle of data by capturing its history from start to finish across all data sources, processes, and systems with details of each activity at a granular attribute level.

Master Data Management

As far back as I remember, it has always been a challenge to reconcile, integrate, and manage master data. **Master Data Management** creates a 360-degree trusted view of scattered master data. Master data, such as that of customers, products, and employees, is consolidated with the intention of generating insights leading to business growth. Master Data Management requires a mature level of data modeling that evaluates several data identifiers, business rules, and data designs across different systems. Metaphorically, you can envision a keyring with an enormous number of keys. Each key represents a unique master data identity, such as a customer. The keyring represents the necessary data harmonization that is realized via Master Data Management across all systems that manage that customer. This is a huge undertaking that requires a mature level of Data Governance and includes Data Quality analysis and data cleansing, such as data deduplication, data standards, and business rules validation.

Metadata Management

The high-quality collection, integration, storage, and standardization of metadata comprise Metadata Management. Metadata, like data, needs to be accurate, complete, and reliable. Data must be mapped with business semantics to make it easy to discover, understand, and use. Metadata needs to be collected by following established interoperability standards across all tools, processes, and system touchpoints to build knowledge surrounding the data. There are four types of metadata that must be actively collected:

- **Technical metadata**: Data structure based on where data resides. This includes details such as source, schema, file, table, and attributes.

- **Business metadata**: Provides business context to the data. This includes the capturing of business terms, synonyms, descriptions, data domains, and rules surrounding the business language and data use in an organization.

- **Operational metadata**: Metadata produced in a transactional setting, such as during runtime execution. This includes data processing-based activities, such as pipeline execution, with details such as start and end time, data scope, owner, job status, logs, and error messages.

- **Social metadata**: Social-based activities surrounding data. This metadata type has become more prominent as it focuses on putting yourself in a customer's shoes and understanding what they care about. What are their interests and social patterns? An example includes statistics that track the number of times data was accessed, by how many users, and for what reason.

Metadata Management is the underpinning of Data Fabric's knowledge layer. Let's discuss this layer in the next section.

Knowledge layer

The **knowledge layer** manages semantics, knowledge, relationships, data, and different types of metadata. It is one of the differentiating qualities of Data Fabric design when compared to other data architecture approaches. Metadata is managed and collected across the entire data ecosystem. A multitude of data relationships is managed across a variety of data and metadata types (technical, business, operational, and social) with the objective of deriving knowledge with an accurate, complete metadata view. The underlying technology is typically a knowledge graph.

The right balance of automation and business domain input is necessary in the knowledge layer. At the end of the day, technology can automate repetitive tasks, classify sensitive data, map business terms, and infer relationships, but it cannot generate *tribal knowledge*. Business context needs to be captured and modeled effectively in the form of ontologies, taxonomies, and other business data models that are then reflected as part of the knowledge layer. Intelligent agents proactively monitor and analyze technical, business, operational, and social metadata to derive insights and take action with the goal of driving operational improvements. This is where **active metadata** plays a major role. According to Gartner, "*Active metadata is the continuous analysis of all available users, data management, systems/infrastructure and data governance experience reports to determine the alignment and exception cases between data as designed versus actual experience*" (`https://www.gartner.com/document/4004082?ref=XMLDELV`).

Let's take a look at what active metadata is.

Moving from passive to active metadata

To understand active metadata, you must first understand what defines passive metadata and why it's insufficient on its own to handle today's big data demands.

Passive metadata

Passive metadata is the basic entry point of Metadata Management. **Data catalogs** are the de facto tools used to capture metadata about data. Basic metadata collection includes technical and business metadata, such as business descriptions. Passive metadata primarily relies on a human in the loop, such as a *data steward*, to manually manage and enter metadata via a tool such as a data catalog. While a data catalog is a great tool, it requires the right processes, automation, and surrounding ecosystem to scale and address business needs. Primarily relying on a human in the loop for Metadata Management creates a major dependency on the availability of data stewards in executing metadata curation activities. These efforts are labor intensive and can take months or years to complete. Even if initially successful, metadata will quickly get out of sync and become stale. This will diminish the value and trust in the metadata.

Another point here is metadata is continuously generated by different tools and typically sits in silos somewhere without any context or connection to the rest of the data ecosystem, which is another example of passive metadata. Tribal knowledge that comes from a human in the loop will always be critical. However, it needs to be combined with automation along with other advanced technologies, such as machine learning, AI, and graph databases.

Active metadata

Now that you understand passive metadata, let's dive into active metadata. It's all about the word *active*. It's focused on acting on the findings from passive metadata. **Active metadata** contains insights about passive metadata, such as Data Quality score, inferred relationships, and usage information. Active metadata processes and analyzes passive metadata (technical, business, operational, and social) to identify trends and patterns. It uses technologies such as machine learning and AI as part of a recommendation engine to suggest improvements and more efficient ways of executing data management. It can also help with discovering new relationships leveraging graph databases. Examples of active metadata in action are failed data pipelines leading to incomplete data where downstream consumption is automatically disabled, and where a dataset with high quality is suggested instead of the current one with low Data Quality, or if Service Level Agreements are not met by a data producer. Alerts that generate action by both data producer and consumer represents active metadata.

Event-based model

The missing link in the active Metadata Management story is the ability to manage metadata live and have the latest and greatest metadata. To drive automation, an event-based model that manages metadata and triggers instantaneous updates for metadata is necessary. This facilitates metadata collection, integration, and storage being completed in a near real-time manner and distributed and processed by subscribed tools and services. This is required in order to achieve *active metadata* in a Data Fabric design.

Let's review the next Data Fabric building block, Data Integration.

Data Integration

Data interoperability permits data to be shared and used across different data systems. As data is structured in different ways and formats across different systems, it's necessary to establish Data Integration standards to enable the ease of data exchange and data sharing. A semantic understanding needs to be coupled with Data Integration in order to enable data discovery, understanding, and use. This helps achieve data interoperability. Another aspect of Data Integration in Data Fabric design is leveraging DataOps best practices and principles as part of the development cycle. This will be discussed in more depth in *Chapter 4, Introducing DataOps*.

Data ingestion

Data ingestion is the process of consuming data from a source and storing it in a target system. Diverse data ingestion approaches and source interfaces need to be supported in a Data Fabric design that aligns with the business strategy. A traditional approach is batch processing, where data is grouped in large volumes and then processed later based on a schedule. Batch processing usually takes place during offline hours and does not impact system performance, although it can be done ad hoc. It also offers the opportunity to correct data integrity issues before they escalate. Another approach is real-time processing, such as in streaming technology. As transactions or events occur, data is immediately processed in response to an event. Real-time data is processed as smaller data chunks containing the latest data. However, data processing requires intricate management to correlate, group, and map data with the right level of context, state, and completeness. It also requires applying a different Data Governance approach when compared to batch processing. These are all factors that a Data Fabric design considers.

Data transformation

Data transformation takes a source data model that is different from a target data model and reformats the source data model to fit the target model. Typically, **extract, transform, and load** (**ETL**) is the traditional approach to achieve this. This is largely applied in **relational database management systems** (**RDBMSs**) that require data to be loaded based on the specifications of a target data model, as opposed to NoSQL systems, where the **extract, load, and transform** (**ELT**) approach is executed. In ELT, there isn't a prerequisite of modeling data to fit a target model like in data lake. Due to the inherent nature of NoSQL databases, the structure is more flexible. Another data transformation approach is programmatic, where concepts such as managing **data as code** apply. When it comes to data transformation, a variety of data formats, such as structured, semi-structured, and unstructured, needs to be processed and handled. Data cleansing is an example of when data transformation will be necessary.

> **What is data as code?**
>
> Data as code represents a set of best practices and principles as part of a DataOps Framework. It focuses on managing data similar to how code is managed where concepts such as version control, continuous integration, continuous delivery and continuous deployment are applied. Policy as code is another variation of Data as code.

Data Integration can be achieved through physical Data Integration, where one or more datasets are moved from a source to a target system using ETL/ELT or programmatic approaches. It can also take place virtually via data virtualization technology that creates a distributed query that unifies data, building a logical view across diverse source systems. This approach does not create a data copy. In a Data Fabric design, all these integration and processing approaches should be supported and integrated to work cohesively with the Data Governance building block, the knowledge layer, and the Self-Service building block.

Self-Service

Self-Service data sharing is a key objective in an organization's digital transformation journey. It's the target of why all the management, rigor, governance, security, and all these controls are in place. To share and deliver trusted data that could be used to create profit. Having access to the right data of high-quality quickly when needed is golden. Data Products are business-ready, reusable data that has been operationalized to achieve high-scale data sharing. The goal of a Data Product is to derive business value from data. In *Chapter 3, Choosing between Data Fabric and Data Mesh*, and *Chapter 7, Designing Data Governance*, we touch more on Data Products. Self-service data moves away from a model that relies on a central IT team to deliver data, to one that democratizes data for a diverse set of use cases with high quality, and governance. Making data easily accessible by technical users such as data engineers, data scientists, software developers or business roles such as business analysts is the end goal.

A Data Fabric design enables technical and business users to quickly search, find, and access the data they need in a Self-Service manner. On the other hand, it also enables data producers to manage data with the right level of protection and security before they open the doors to the data kingdom. The Self-Service building block works proactively and symmetrically with the Data Integration and Data Governance building blocks. This is where the true power of a Data Fabric design is exhibited.

Operational Data Governance models

Data Fabric architecture supports diverse Data Governance models. It has the necessary architecture robustness, components, and capabilities to support three types of models:

- **Centralized Data Governance**: A central Data Governance organization that is accountable for decision-making, data policy management, and enforcement across all business units within an organization. This represents a traditional model.

- **Decentralized Data Governance**: An organization where a business domain is accountable for and manages the Data Governance program. The business domain is accountable only for their domain's decision-making and data policy management and enforcement. They operate independently in terms of data management and Data Governance from the rest of the organization.

- **Federated Data Governance**: In a federated Data Governance model, a federated group of business domain representatives across an organization and governance roles is accountable for global-level concerns and decision-making on Data Governance matters. The federated governance team is not responsible for enforcement or local policy management. Business domains manage their data, local data policies, and enforcement.

Let's summarize what we have covered in this chapter in the next section.

Summary

In this chapter, we defined Data Fabric and its characteristics–highly automated, use case agnostic, Self-Service, strong Data Governance, Privacy, and Protection, supports decentralized, federated, and centralized Data Governance, event and active metadata driven, interoperable, and has a composable architecture. We discussed why this architecture is important, including its ability to effectively address data silos, data democratization, establish trusted fit for purpose data while addressing business and technical needs. We introduced the core building blocks of a Data Fabric design (Data Governance, Data Integration, and Self-Service) and its principles. We reviewed the role and value of Data Fabric's knowledge layer, uses active metadata as the glue that intelligently connects data across an ecosystem. We also defined the various Data Governance models that Data Fabric supports.

In the next chapter, we will dive into the business value that Data Fabric architecture offers. We will discuss how the components that make up the backbone of Data Fabric derive business value that leads to profit and cost savings.

Show Me the Business Value

Business value is measured by the success of profit and revenue. In data management, this translates into the ability to monetize data. Pain points in data management often lead to cost-intensive business operations or the loss of business opportunities, impacting revenue. Data Fabric architecture addresses common complexities in distributed data management, such as data silos, data protection, and data democratization. Data Fabric is able to tackle a diverse set of pain points via its core building blocks – **Data Governance**, **Data Integration**, and **Self-Service**. Each building block has an area of specialty that works cohesively with the other building blocks, leading to higher profits and cost savings.

This chapter defines what digital transformation is, the role of data monetization, and Data Fabric's value proposition. It reviews the benefits of Data Fabric architecture and its applicability across organizations. By the end of this chapter, you will understand the business value behind high-quality, secure, integrated, trusted data being available in a Self-Service manner. You'll get a business point of view of how Data Fabric architecture can help organizations achieve higher profits and drive down operational costs.

In this chapter, we will cover the following topics:

- Digital transformation
- Data monetization
- Data Fabric's value proposition
- Data Fabric for large, medium, and small enterprises

Digital transformation

Digital transformation is the journey in which organizations seek to leverage technology and data to achieve financial returns. This is typically done by using technology to harvest data with automation, quality control, and meaningful insights. The outcome is a modern business model that creates new business opportunities. It offers a value-driven, customer-focused approach. The quicker you execute the data insights derived, the faster you achieve value, whether in the form of profit or operational efficiencies that drive down cost. There is constant pressure to execute rapidly or risk losing business to competitors. Organizations that implement a digital mindset understand the tremendous value and the survival need to become digital. Businesses that do not grasp this are in danger of getting left behind, while digital companies thrive and keep up with a fast-paced market. An article on the MIT Sloan Management Review agrees with this premise.

> *… among companies with over $1 billion in revenues, 24% had digitally savvy boards, and those businesses significantly outperformed others on key metrics — such as revenue growth, return on assets, and market cap growth.*

> **Important note**
>
> *It Pays to Have a Digitally Savvy Board* by P. Weill, T. Apel, S. L. Woerner, and J. S. Banner, *MIT Sloan Management Review 60*, no. 3 (2019): `https://sloanreview.mit.edu/article/it-pays-to-have-a-digitally-savvy-board/`

Digital transformation focuses on the value behind leveraging technology and data, which in turn creates data monetization. Let's understand what composes data monetization.

Data monetization

Organizations leveraging data in their business aim to achieve financial return in the form of revenue or cost savings. They use facts backed by data to set an execution strategy on the who, what, where, when, and why. The business objectives are to derive significant and actionable insights from data leading to monetization. Gartner defines data monetization as follows:

> *…process of using data to obtain quantifiable economic benefit. Internal or indirect methods include using data to make measurable business performance improvements and inform decisions. External or direct methods include data sharing to gain beneficial terms or conditions from business partners, information bartering, selling data outright (via a data broker or independently), or offering information products and services (for example, including information as a value-added component of an existing offering).*

`(https://www.gartner.com/en/information-technology/glossary/data-monetization#:~:text=Data%20Monetization%20refers%20to%20the,performance%20improvements%20and%20inform%20decisions)`

Let's discuss, in the next two sections, data monetization in the form of revenue and cost savings.

Revenue

Data insights are typically focused on the customer, product, or service. To achieve success, businesses today must provide customers with a personalized experience that focuses on their unique needs. It's human nature to want to be heard, recognized, and treated as an individual. This leads to organizations spending a lot of time and money to understand a customer's wants and needs. What kind of customer are they? What are their interests? What are their typical purchases? When do they make these purchases? The goal is to identify patterns that provide a sufficient understanding of their interests to take action ultimately leading to revenue. Another form of revenue can be in the direct monetization from the selling and purchasing of data as part of a data marketplace solution.

Cost savings

Data insights can also be used to achieve cost savings. It can identify non-value-added business processes and expensive operations as candidates to drive process quality improvements. The target is to save on unnecessary costs and find alternative, more efficient ways to execute those business processes. This can be achieved in many ways depending on the industry of an organization. A few examples can entail finding alternative, lower-priced materials to manufacture goods, automating the packaging of goods with machinery, and more efficient delivery routes to cut down on gas. It offers a deeper look, powered by data, into the life cycle of a customer, product, or service to propel lowering costs to achieve increased profits.

The following is a perspective on successful data monetization with a data lens view. It isn't meant to serve as the ultimate financial formula but as a starting point for how to think about it. There are other elements that need to be considered across the technology, people, and process dimensions, for example, having the right business model to achieve data monetization.

Trusted quality data + meaningful insights + action-oriented business plan + high execution speed = profitable data monetization

Let's look at each of the variables that contribute to data monetization:

- **Trusted quality data**: The success of a data monetization strategy starts with governed data that is of high quality or fit for purpose where it can be trusted and used to make high-stake business decisions. This includes ensuring the data has been protected to address regulatory requirements and secured properly. Metrics defined and tracked by businesses typically assume that data is reliable, can be used without ramifications, and is of high quality. As an example, if sales predict an uplift of 40% for future customer purchases, there will need to be a considerable amount of investment in marketing to achieve this. If it turns out that this prediction is made based on untrustworthy data, it can cause the predictions to be inaccurate, thereby leading to huge monetary loss.

- **Meaningful insights**: Meaningful insights rely on trusted, high-quality data. Although, let's apply a slightly different lens. Having a complete and correct data scope provides an accurate data representation that establishes correlations and patterns. Data needs to be integrated across multiple data sources, data formats, and environments with the appropriate logic and speed. Otherwise, the insights won't have a complete picture, resulting in missed business opportunities. Even if the Data Integration approach was correct, but delivered late, then it doesn't produce business value due to the flaws in the Data Integration approach.

- **Action-oriented business plan**: A well-thought-out business plan that acts on the insights with the right level of priority aligned to business objectives is necessary. It is important to measure the success of your execution with data insights. Some example questions to drive measurement are: Was a marketing email successful in leading to a customer purchase? If yes, by how much? You will see a pattern emerging on the business value that trusted governed data with quality offers.

- **High execution speed**: Rapid speed in getting access to data to execute is key. There is a major dependency on data democratization. Getting the data needed, exactly when needed, is required. If you request access to data from a central IT team and that request isn't fulfilled until weeks or months later, it becomes a barrier to having a competitive advantage. Time is money and having Self-Service access to data is vital.

These variables create some of the elements in a successful data monetization formula. The critical success factors are governed, trusted, and rapid data access.

In the next section, let's take a deeper look at how Data Fabric's building blocks create business value, thereby creating a recipe for data monetization.

Data Fabric's value proposition

Each building block in a Data Fabric architecture (Data Governance and its knowledge layer, Data Integration, and Self-Service) introduced in *Chapter 1* offers a monumental amount of value from its sophisticated approach and focus. *Figure 2.1* provides a summary of the value statements in Data Fabric's building blocks.

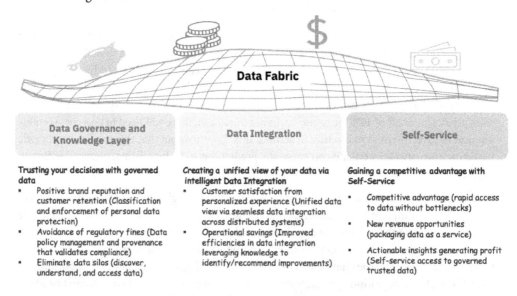

Figure 2.1 – Business value via Data Fabric's building blocks

Let's discuss the value behind the Data Governance building block.

Trusting your decisions with governed data

Data Governance is a staple in achieving a successful data monetization strategy. In Data Fabric architecture, Data Governance is focused on producing high-quality, trusted, secure, discoverable, and understandable data. Data consumers rely on trusted data to address their business needs.

Let's review the value proposition across the Data Governance pillars.

Protecting and securing critical data

Data protection concentrates on preventing data breaches, which in turn save large sums of money for businesses. It involves preventing unauthorized access to data deemed as sensitive or private, based on established government laws as in Europe's **General Data Protection Regulation (GDPR)** and the United States' **California Consumer Privacy Act (CCPA)** and **Health Insurance Portability and Accountability Act (HIPAA)**. An organization must take the necessary precautions to handle sensitive, personal, and confidential data with encryption and enable access to only authorized users. Both data at rest and data in flight need to be addressed. Data protection typically involves a series of steps, such as data classification, data privacy, security policies, restricted access control, user authentication, and data masking.

There are consequences for organizations that do not address data protection and privacy. Implementing data privacy efforts minimizes the risk of data breaches, therefore avoiding regulatory fines and negative brand impact. Millions of dollars can be saved by organizations that comply. Several companies have already experienced these massive losses in fines, which solidifies the criticality of properly protecting and securing data.

GDPR has imposed extremely high fines on the following companies:

- Amazon was fined 746 million euros: `https://www.cnbc.com/2021/07/30/amazon-hit-with-fine-by-eu-privacy-watchdog-.html`

- WhatsApp was fined 225 million euros: `https://www.bbc.com/news/technology-59348921`

- Google was fined 50 million euros: `https://www.nytimes.com/2019/01/21/technology/google-europe-gdpr-fine.html`

Apart from the high dollar amount in fines by government agencies, there is another loss – the brand's reputation. Forbes stated the following:

> *The same way people now check how a company's actions impact the environment and seek to buy from more sustainable brands, the world is taking a closer look at companies' privacy policies to learn about the values behind the name.*

(https://www.forbes.com/sites/forbestechcouncil/2021/10/07/
why-data-privacy-is-good-for-business-online-privacy-as-a-
branding-imperative/?sh=2ef6dabb297e)

A company's reputation may be negatively impacted if the proper steps to implement and enforce data privacy policies are not taken. It can lead to lost customer opportunities, thereby impacting revenue. Customers and employees prefer to work with/for companies that openly advocate and implement data privacy protection as opposed to companies that have experienced data breaches. Addressing data privacy and protection with the right level of control and security is critical in the journey of achieving data monetization. It's counter-intuitive to achieve high-profitable revenue and then lose it by paying regulatory fines or later losing profit due to negative brand impact.

Driving better business outcomes with quality data

It is important to ensure that the data used to make critical business decisions is of high quality and reliable. Business decisions that lead to increased productivity and customer success are the objectives. A major assumption made by businesses when deriving insights is that the data is trustworthy. Data is assumed to be of high quality and free from data defects. As a result, organizations invest large sums of money in strategic execution plans. This includes the generation of marketing campaigns, customer lifetime value analysis, and process improvement initiatives all in an effort to increase revenue and cut down operational costs. The downfall here is the incorrect assumption of high Data Quality without proper Data Governance in place focused on ensuring Data Quality.

Achieving governed and trusted data can lead to a significant increase in revenue for an organization. In the retail industry, a marketing campaign can more accurately target the relevant customers. In the banking industry, you can take calculated risks based on a customer's profile, credit information, and loan history. However, if these decisions are made based on incorrect data, then an organization can lose millions of dollars in failed marketing campaigns or potentially defaulted bank loans. Gartner's research in 2018 (https://www.gartner.com/smarterwithgartner/how-to-create-a-business-case-for-data-quality-improvement) indicated that organizations lose on average $15 million per year due to poor Data Quality.

Oftentimes, to put into context the impact of bad Data Quality in data science, the 80/20 rule is mentioned. Data scientists spend 80% of their time doing data cleansing and only 20% doing data analytics. Data-driven roles spend laborious time on non-value-added activities such as scrap and rework. According to *Larry English*, in his book *Information Quality Applied*, the cost of poor-quality information is defined by the wasted time and effort spent to correct data and control redundancy. Larry English was a thought leader in information quality, often referred to as the *father of information quality*. I had the privilege of learning from him a core set of Data Quality management principles as part of the **Total Information Quality Management** (**TIQM**) framework. In the TIQM framework, he calls data scrap and rework *Information Muda*. The premise is that spending time correcting bad-quality data does not provide any business value. Instead, quality needs to be designed directly into the data processes and pipelines that create, update, and deliver data to data consumers. This provides the means necessary for data consumers to take the correct course of action and make business decisions leading to business value.

Validating compliance with regulatory laws via Data Lineage

Similar to Data Quality, Data Lineage establishes data trust. It provides a view into the life cycle of data from both business and technical perspectives. It provides answers to data provenance questions – where did the data come from? What systems, applications, and processes did it flow through? What data transformations were applied along the way? Who accesses the data? With a large number of distributed systems and roles that work on data, it's important to keep track of the source of data and its journey across all data storage systems, including transformations applied. Data Lineage is absolutely necessary for compliance and traceability. It provides the audit trail needed for addressing regulatory compliance in data protection and privacy. Knowing who accessed the data, when, how, and for what purpose needs to be provided as proof to showcase compliance and avoid regulatory fines. Data Lineage is also essential in resolving Data Quality issues. It can reveal inefficient and non-value-added operations, such as multiple data providers creating the same data, leading to data duplication. It traces the life cycle of data to drive proper data cleansing and process quality improvement.

Addressing data silos with discoverable and understandable data

Metadata Management enables data discovery, data understanding, and the usability and reusability of data. It provides a view into an organization's data inventory with business context to enable its use. Metadata Management requires the curation of data by associating and producing relevant context to data. Metadata Management solves the pain point of *dark data* or data that is hoarded across different systems. There are huge costs associated with the lack of metadata or low-quality metadata. According to Verge, missing or incorrect metadata is a huge problem in the music industry. It has led to a growing number of artists not getting paid royalties for their music, leading to the loss of billions of dollars (`https://www.theverge.com/2019/5/29/18531476/music-industry-song-royalties-metadata-credit-problems`).

Master Data Management (**MDM**) is a subset of Metadata Management and intersects with Data Quality and Data Integration. High-quality data that creates a 360-degree view of a customer and other master data leads to better insights and increased revenue. As an example for customer data, product incentives could be offered to a client based on their purchase history, which requires the unification of data across distributed systems.

Superior decision-making with a holistic view across complex relationships

The knowledge layer in Data Fabric architecture creates a holistic data view and manages complex data relationships. It offers tremendous powers in advanced decision-making. To showcase the value of the knowledge layer, let's use the human brain as an analogy. Let's say a close friend of mine really enjoys Italian cuisine. I recently went to an amazing Italian restaurant in the suburbs of my city, and knowing my friend enjoys Italian cuisine, I intuitively recommended the restaurant to her. After, I learn that she enjoys city life, and typically ventures into the city each weekend. Later that day, I speak to another friend of mine who raves about an Italian restaurant located in the city. I immediately connect the two situations and recommend the Italian restaurant located in the city to my city-loving friend. The inherent intelligence a human brain has to create context, make associations, and create relationships based on knowledge derived from events, subjects, entities, and contexts and create recommendations based on this is how a knowledge layer works.

Now, imagine applying this level of power to organizational data, business processes, technologies, tools and managing Data Governance with high-scale data sharing. This is where the biggest value is generated by Data Fabric architecture.

Let's discuss the business value behind Data Fabric's Data Integration building block.

Creating a unified view of your data with intelligent Data Integration

A high degree of data interoperability is absolutely necessary to facilitate the ease of data sharing across complex architectures today. Efficient Data Integration helps achieve data interoperability, and data interoperability allows successful data enablement. Businesses need data for monetization. Before data can be used, it needs to be ingested and integrated to achieve data interoperability. Data Integration is the act of combining data across several source systems into a target unified view. The intention behind Data Integration is to extract actionable insights leading to data monetization.

Today's data landscape consists of several data systems spread across different environments. There is major complexity in the processing of data, such as understanding data models across data systems, mapping business definitions, and applying the correct business rules in the right sequence. A superior Data Integration approach is needed, otherwise poor integration leads to unnecessary costs. According to the *2019 State of Ecosystem and Application Integration Report*, poor integration costs businesses half a million dollars every year. A high-performing Data Integration approach is imperative in providing the speed and unification of data required. It must address the data silo barricades and accomplish this productively. If you compare this to the objectives of data monetization, Data Integration is a must in achieving a custom experience for the customer. There are many factors that drive how successful you are in achieving data monetization, a major one being the Data Integration approach you take.

Now that we have covered the business value for Data Fabric's Data Integration building block, let's take a look at Self-Service.

Gaining a competitive advantage with Self-Service

Data democratization breaks down data silos by enabling rapid Self-Service data access with automation. It is the opposite of a traditional data access approach, which relies on a central **Information Technology (IT)** team. In a traditional approach, it is often viewed as an IT team holding the keys to the data kingdom. There typically is a long turnaround period from the time data access is requested by a data consumer to the time it takes to deliver the data. Traditional data access processes contain many manual tasks, require building custom solutions for every data request, lack proper business controls that address data security, and lead to unacceptable turnaround times.

Self-Service focuses on delivering data quickly with security and high quality. Self-Service data access offers data consumers more control over the data they need and enables collaboration with data producers and other data consumers. It addresses the typical bottlenecks and quality issues in a traditional approach. Self-Service empowers data consumers of both business and technical roles to access and consume data for their needs. It is a systematic approach to data delivery and enables quality insights, leading to customer success. Some of its key principles are automation, ease of discovery, data reusability, standardization, governed access, effective collaboration, and quick data delivery. It offers an enormous amount of value by providing the data needed when it's needed. An example of Self-Service in action is via a data marketplace tool. It facilitates data sharing between data producers and data consumers, typically at a cost. It facilitates sharing data as a service. According to a paper by Accenture, IoT data marketplaces can potentially make $3.6 trillion by 2030 (`https://www.accenture.com/_acnmedia/PDF-85/Accenture-Western-Digital-Value-of-Data-Dawn-of-the-Data-Marketplace.pdf`).

Data Fabric for large, medium, and small enterprises

Large, medium, and small enterprises can derive value from Data Fabric architecture. It offers deep Data Governance that protects and secures data with high-quality, embedded automation, efficient Data Integration, and Self-Service. It is use case agnostic across a diverse set of industries. A Data Fabric solution offers a robust and scalable platform that can handle the evolving needs of businesses and manage data efficiently and effectively. A Data Fabric architecture-based solution is flexible and composable, which offers organizations the ability to start small or big depending on the business priorities of an organization. There can be different entry points for organizations of all sizes, which we'll discuss in the following subsections.

Large enterprise organizations

Typically, large enterprises have complex architectures, commonly referred to as spaghetti architectures or *big balls of mud*. They are extremely complex and hard to understand and have a large number of systems, data environments, several business domains with intricate processes that are dependent on each other, and data silos. There is a huge motivation for large organizations to implement a Data Fabric architecture solution as it excels in addressing many of the pain points around data silos, data democratization, regulatory compliance, and enables businesses to make decisions they can trust.

Small and medium-sized businesses

Small and Medium-Sized Businesses (SMBs) typically have constraints in terms of the amount of funding and resources that can be invested. The same level of architecture complexity is not exhibited when compared to large enterprise organizations. However, it's important for SMBs to understand the critical importance of data analytics and how Data Fabric architecture can offer a path to achieve this with the right level of control and Self-Service. This can lead to a differentiating role in the market, providing a competitive advantage, especially as SMBs stack up against large enterprises. Getting the right data quickly is even more critical for SMBs. It can be what decides whether an SMB stays in business. Poor business decisions in SMBs may not be as easy to recover from when compared to large enterprises. SMBs can prioritize the areas of focus based on their organizational needs. Data Fabric is a composable design that enables SMBs to start small and evolve as they mature in their digital journey.

Let's summarize what we've learned in this chapter.

Summary

In this chapter, we discussed the business value of Data Fabric architecture. We spoke about the value proposition of each of Data Fabric's building blocks. We have also posited Data Fabric architecture as a strategic approach to successful data monetization. The Data Fabric building blocks – Data Governance, Data Integration, and Self-Service – work together to deliver data that can lead to highly profitable revenue and operational efficiency. We have also reviewed real-life examples showcasing millions of dollars lost by organizations not having the right data architecture and strategy in place. We have also discussed the value Data Fabric architecture can offer large, medium-sized, and small organizations. Data Fabric can substantially provide a huge return on investment across organizations focused on digital transformation.

In the next chapter, we will review Data Mesh architecture and discuss how the Data Fabric and Data Mesh architectures are complementary.

Part 2: Complementary Data Management Approaches and Strategies

For organizations to become data driven, they need a comprehensive view of other complementary data management approaches that can accelerate their business value.

This second part consists of three chapters that introduce Data Mesh and DataOps as complementary data management approaches. Each data management approach is discussed with a dedicated focus on explaining how it can be applied together with Data Fabric architecture. *Part 2* ends by going through an approach to creating a data strategy document.

By the end of this part, you will have a view of how trending data management approaches such as Data Mesh and DataOps could be leveraged together with Data Fabric architecture. You will also have a reference on how to create a data strategy document.

This part comprises the following chapters:

- *Chapter 3, Choosing between Data Fabric and Data Mesh*
- *Chapter 4, Introducing DataOps*
- *Chapter 5, Building a Data Strategy*

3

Choosing between Data Fabric and Data Mesh

Just like Data Fabric, Data Mesh has become a buzzword in data management. Data Mesh focuses on principles such as managing data as a product, Self-Service, and providing a federated model at the organization and Data Governance level. The premise is to dismantle the concept of relying on central teams for data operations and move away from centralized data management to a decentralized approach. The Data Fabric and Data Mesh architecture approaches are often confused or referred to interchangeably; however, they are two separate design concepts. The question often asked is, do we need to choose one over the other? The answer is that they are complementary, and Data Fabric can be used together with Data Mesh.

In this chapter, we'll focus on discussing how Data Fabric and Data Mesh share similar best practices and principles, and we'll look at where they differ. I'll provide a perspective of Data Mesh based on *Zhamak Dehghani*'s book titled *Data Mesh*. Dehghani is responsible for developing the Data Mesh design.

By the end of this chapter, you'll understand the complementary nature of Data Fabric and Data Mesh. You'll have seen how a Data Fabric's use-case-agnostic and flexible, active, metadata-driven approach accelerates the realization of a Data Mesh design.

This chapter will cover the following topics:

- Introducing Data Mesh
- Comparing Data Fabric and Data Mesh
- Data Fabric and Data Mesh's friendship

Introducing Data Mesh

Data Mesh is a design concept based on federated data and business domains. It applies product management thinking to data management with the outcome being *Data Products*. It's technology agnostic and calls for a domain-centric organization with federated Data Governance. Data management and local governance are owned and implemented at the business domain level. Cross-cutting business domain policies are created by a federated governance team at a global level but executed locally by the domain teams. The business pain points Data Mesh focuses on are resolving organizational bottlenecks, data silos, data swamps, and alignment between business domains and Data Governance teams. The following is Dehghani's definition of Data Mesh:

> *Data mesh is a decentralized sociotechnical approach to share, access, and manage analytical data in complex and large-scale environments within or across organizations.*

In her book, Dehghani positions Data Mesh as having four core principles:

- Domain ownership

- Data as a product

- Self-Serve data platform

- Federated computational governance

In the following subsections, you will find a summary of each of these principles.

Domain ownership

A business domain owns its data and the sharing of data for the entire organization. This principle is focused on eliminating the central IT team bottlenecks to facilitate scaling. It also addresses the lack of business ownership necessary to drive change and the required expertise to accurately understand and use data. This style of data ownership is inspired by the best practices introduced in **domain-driven design**, a software design approach. Data ownership exists across three areas:

- Source

- Consumer level

- Aggregation of data across the first two areas

Domain ownership draws a boundary between analytical and operational data storage. However, both analytical and operational data are owned by the same business domain with a clear emphasis on data interoperability across diverse business domains.

Data as a product

Managing data as a product is focused on addressing data silos and quality challenges. It advocates a culture where Data Product owners are accountable for creating value-driven Data Products with the end consumer in mind. There is a heavy emphasis on reusability, data interoperability, and data trust. You can compare this concept to software product management except this applies a product management approach to managing the life cycle of data, instead of software. The focus is to create high-quality Data Products that address a customer's data access pain points. The following characteristics define a Data Mesh Data Product: discoverability, addressability, understandability, trustworthiness, native accessibility, interoperability, independently valuable, and security.

Self-Serve data platform

A data democratization platform that enables data consumers to easily access and consume Data Products with speed is the focus of this principle. It enables business domains to create and manage the life cycle of Data Products with automation based on federated computational governance. Domain product owners create reusable, high-quality Data Products that are discoverable, governed, and accessible via this platform. Technologies are applied to enable managing Data Products in a distributed manner as opposed to taking a central management approach. A Self-Service data platform hides the complexities of building, maintaining, and sharing Data Products with high interoperability.

Federated computational governance

Federated Data Governance is managed at two levels: the global level and the business domain level. Federated computational governance embeds governance in an automated manner directly into Data Products leveraging the Self-Serve platform. At an organization level, the emphasis is for the majority of the data policies to be defined at a business domain level, while a smaller portion of the data policies that are cross-functional are defined at a global level.

Now that we have provided a perspective on the four principles of Data Mesh, let's compare and contrast Data Fabric and Data Mesh.

Comparing Data Fabric and Data Mesh

Data management best practices and principles have evolved over decades to address common pain points faced today with managing data and getting access to data quickly with the goal of achieving high business value. Organizations have learned the value of data through pain, failure, and revenue loss. This has led to an organizational culture change where data is no longer treated as a by-product but instead as a valuable asset, and data is seen as a product that needs to be treated with care, strategic planning, and much attention to ultimately achieve data monetization.

The reality is that both the Data Fabric and Data Mesh designs have evolved from lessons learned from previous data management approaches. They are based on a collection of best practices and principles to execute superior data management and achieve high-scale data sharing. They address the growing needs and complexities of data management across the four Vs – volume, velocity, variety, and veracity. Both the Data Fabric and Data Mesh designs propose a new approach with different execution, although complementary, styles.

Let's revisit Data Fabric's definition and compare it to Data Mesh's definition.

Objectives

Here is a recap of Data Fabric's definition from *Chapter 1, Introducing Data Fabric*:

Data Fabric is a distributed and composable architecture that is metadata- and event-driven. It's use case agnostic and excels in managing and governing distributed data. It integrates dispersed data with automation, strong data governance, protection, and security. Data Fabric focuses on the self-service delivery of governed data.

> **Important note**
>
> It is important to note that Data Fabric has been defined in this book as excelling in decentralized and federated data management but can also support centralized data management. There are other points of view in the market that define Data Fabric as a centralized data management solution. While I agree that a Data Fabric design can support a central data management approach to Data Governance, it doesn't require data to be stored in a central data system. It supports distributed data across systems, tools, and environments. Data Fabric's power is in the very fact that it embraces both centralized and decentralized data management via its active metadata-driven design.

Data Fabric and Data Mesh have similarities in their definitions. They both call out the need for rapid data access, high-quality data, and Data Governance. They acknowledge today's data complexities and propose a decentralized data management approach. Now that we have got an idea of both Data Fabric's and Data Mesh's definitions, let's compare and contrast them across four groups of data management characteristics:

- Organizational structure
- Data management style
- Data platform
- Data Governance

Let's look at each of them in detail.

Organizational structure

Data Fabric has a strong culture that's heavily focused on Metadata Management. It's all about the metadata, and more specifically, active metadata to achieve value-driven data management. A Data Fabric culture advocates for high-quality, complete, and active metadata. Data Fabric does not require a specific organizational structure as in Data Mesh. A Data Fabric design is heavily focused on driving success via the core characteristics discussed in *Chapter 1, Introducing Data Fabric*, which are also listed here:

- Highly automated

- Use case agnostic

- Self-Service

- Strong Data Governance, Privacy, and Protection

- Supports decentralized and federated data management

- Event driven

- Active metadata driven

- Interoperable

- Composable architecture

Data Fabric makes some assumptions, such as there needs to be a specific business entity within an organization that drives the core Data Fabric attributes, such as a Data Governance authority. Its focus is to support diverse organizational models, whether top down, bottom up, or in between. Data Fabric supports a federated, decentralized, or centralized organization. To participate in Data Fabric, metadata is contributed in an automated manner and knowledge is populated from it to propel data management. Data Fabric is different from a Data Mesh design in that it supports decentralized, federated, and centralized organizations. Data Fabric's objectives are to help an organization to evolve to a more mature level of data management by leveraging active metadata, which is a core prerequisite.

Data Mesh's core prerequisite is a federated organization of business domains. It derives its inspiration from software development's domain-driven design. It does not support the centralization of data teams as Data Fabric does. It moves away from this organizational structure to instead have **Data Product owners** that own and manage Data Products within their respective business domains. To participate and fully execute a Data Mesh, you must be organized as a federated organization and follow Data Mesh's four principles:

- Domain ownership

- Data as a product

- Self-serve data platform

- Federated computational governance

Let's move on to the data management style.

Data management style

Both Data Fabric and Data Mesh support decentralized data management with an emphasis on Self-Service data access that addresses the complexities of managing data across distributed data systems in hybrid and multi-cloud environments. Both recognize the difficulties of today's data architectures and the importance of managing data as a product and not a by-product. In the following list, I provide a perspective on Data Mesh's usability characteristics, which are also characteristics of and supported by Data Fabric:

- **Discoverability**: The ability to search for and find data by leveraging metadata

- **Addressability**: The ability to access data uniquely irrespective of data or metadata updates

- **Understandability**: Marrying data with the right level of semantics and context to provide business understanding

- **Trustworthiness**: Data that can be trusted and relied upon to make critical business decisions (for example, Data Quality and Data Lineage)

- **Native accessibility**: Support different modes of data access tailored to the types of data consumers and their use cases, such as in data science, data engineering, or business intelligence

- **Interoperability**: Facilitate the ease of data use and data unification across a distributed architecture

- **Independently valuable**: Must provide business value and have the right level of Data Governance to avoid risks

Data Fabric and Data Mesh do not share the same execution style in managing data. This is where they differ. Data Fabric offers flexibility in its execution, making it adaptable to a diverse set of use cases. It uses an active metadata approach to collect metadata across an organization that manages data using various data management styles and operational models. In the case of Data Mesh, the implementation details are left to the technology and it implements more of a process and organization focus on how data needs to be created and managed. In Data Mesh, data is created and managed by the domain. Where Data Fabric and Data Mesh complement each other is in their shared emphasis on connecting data across distributed business domains via data interoperability standards to achieve data trust and data security. Both heavily advocate data sharing.

Data Fabric acknowledges the strengths and weaknesses of both decentralized and centralized data management and can support both, as opposed to Data Mesh, which only supports decentralized data management. Data discoverability is key in both approaches. However, Data Mesh advocates for this to be done closer to the source, whereas Data Fabric has the means to make this real via its metadata approach and is flexible in the location. Discovery in Data Fabric can take place at the source like in Data Mesh, or in a system such as a data warehouse, data lake, or data lakehouse.

DataOps is another superhero data framework that both Data Fabric and Data Mesh recognize as critical to successful data operations. It identifies the need for managing data as code to drive automation, scalability, and quality. Data Fabric focuses on end-to-end data orchestration to achieve a single-pane-of-glass view while connecting the dots between data relationships via its knowledge layer. Data Fabric can orchestrate data across an entire ecosystem or at a granular level in a Data Product as Data Mesh calls for. Data Mesh applies DataOps and policies as code principles in the development of domain-based Data Products. Both Data Fabric and Data Mesh focus on data interoperability standards to connect data.

Data platform

Self-Service data access is a staple in achieving data monetization. The quicker you can access and use high-quality data, the more competitive advantage you will have. Both Data Fabric and Data Mesh agree on the need for a Self-Service data platform. They align on the capabilities necessary, such as strong Data Governance; seamless, automated data; Metadata Management; and enabling the data discovery, data understanding, data access, and data consumption of high-quality data. The data platform must offer DataOps features to accomplish data observability, CI/CD, automated testing, and automated deployment, as well as managing data and infrastructure as code. It must also offer an easy and intuitive experience in the management of Data Products and data delivery, such as the use of APIs that abstract the Data Integration complexities. It must also offer capabilities with a focus on Data Products, data sharing, data interoperability standards, and achieving scalability. The data platform needs a Self-Service data architecture that can search, understand, access, govern, and deliver data with automation.

Where Data Mesh differs from Data Fabric is that it has fixed requirements for the Self-Service platform focused on organizing and managing Data Products by business domain. Another difference is Data Fabric supports managing data as an asset and as a product. A Data Product can be composed of assets that have been governed and managed in a Data Fabric architecture. Data Fabric does not have these fixed requirements, although it inherently supports isolating data and Data Governance enforcement via metadata by business domain. You can think of a Data Mesh Self-Service data platform as supporting separate, independent companies (business domains), although the key criteria are that it does not create data silos and attains data sharing across these companies in a secure, quick, and easy manner. In Data Mesh, Data Products are created and managed by federated business domains and a data platform requires capabilities that enable data and policy federation. This is where a Data Fabric solution can also address Data Mesh's requirements.

Data Governance

Data Governance is recognized as critical to both Data Fabric and Data Mesh. Success cannot be achieved without the right level of Data Governance and automation. Today, there are three styles of Data Governance:

- Top down, which is centralized Data Governance

- Bottom up, which is decentralized Data Governance
- Federated Data Governance

Data Fabric supports all three Data Governance styles, as opposed to Data Mesh, which focuses on a federated model. Data Fabric and Data Mesh agree that data must be governed with a high degree of automation to achieve scalability by embedding it directly as part of data operations. Both approaches have a set of key principles necessary to realize data interoperability and enable effective data communication. In the case of Data Mesh, Data Products are created independently by each business domain, although it must enable cross-domain data communication in a secure and private manner.

Both approaches, Data Fabric and Data Mesh, agree on the need for **feedback loops and metrics** and **leverage points**. Feedback loops and metrics provide perspective and the ability to influence the quality of data over time. The intention is to drive action based on the knowledge collected, which aligns back with the premise of Data Fabric's active metadata characteristic. The objective is to produce high-quality data with feedback from end users by collecting social and operational metadata. Data Fabric leverages a number of technologies, tools, and methods to achieve automated Data Governance with mature access control, data security, and data observability.

Data Mesh is more specific in comparison to Data Fabric in how Data Governance should be executed. It requires a federated Data Governance team made up of a cross-functional team, **data platform representatives**, and data privacy and security SMEs. This is one of the major differentiators of a Data Mesh design. Data Governance policies that have a cross-domain impact are defined at a global level. However, the execution takes place locally by the business domains. Automated enforcement is the responsibility of the data platform and requires Data Governance to take place close to the originating source. The business domains are responsible for implementing Data Governance based on the set global data rules for a Data Product, such as Data Quality.

Data Mesh distinguishes between the role of a data steward focused on enterprise data in traditional Data Governance and a role bound to the business domain level, referred to as the Data Product owner. Data Fabric supports traditional Data Governance and Data Mesh-specific roles. Data Mesh advocates for the roles of **data platform product owner** and **platform architect** to be a part of a federated team responsible for Data Governance. These roles aren't new but rather existing roles such as software product owners and technical architects. What is different in Data Mesh from traditional Data Governance is that these roles execute Data Governance at the Data Product level within a business domain. They must also package a Data Product with code, metadata, data, and policies.

In the following table, you will find a summary of the comparison between Data Fabric and Data Mesh design:

Pillars	Data Fabric	Data Mesh
Organizational Structure	• Strong metadata-driven culture • Supports federated, decentralized, and centralized organizations • Participation by taking an active metadata approach	• Organizationally driven following a federated model • Does not support a centralized-based model for data management • Participation via business domain ownership and accountability in data management
Data Management Style	• Primarily Process, and Technology driven. Flexible in supporting the People dimension • Supports both decentralized and centralized data architectures • Manages data as an asset, data as a product, and derived data assets, i.e., machine learning models and dashboards • Leverages DataOps, active metadata, AI/machine learning, and recommendation engines to drive intelligent data management and Data Integration • Self-Service data consumption • Strong emphasis on metadata maturity, automation, Data Governance, and data interoperability • Data Observability with an emphasis on automation	• People, and Process driven • Supports decentralized data management (moves away from centralized data platforms) • Manages data as a product at a business domain level autonomously • Technology agnostic • Self-Service Data Product consumption • Data Governance embedded in a Data Product • Strong data interoperability • Leverages DataOps

Pillars	Data Fabric	Data Mesh
Data Platform	• Distributed data architecture with a composable set of tools, technologies, and data storage solutions • Supports decentralized, federated, and centralized data management • Manages data assets and a Data Product's life cycle • Capabilities focus on DataOps-based Data Integration, Self-Service, active metadata, and Data Governance • Capabilities support use cases or support data/Metadata Management	• Distributed data architecture • Enables business domain teams to create and share Data Products in a Self-Service manner • Capabilities support a federated organization, automated Data Governance, and DataOps • Life cycle of Data Products is managed • Data product architecture consists of one unit (code, data, metadata, and policies)
Data Governance	• Supports decentralized, federated, and centralized Data Governance • Supports traditional Data Governance roles such as data stewards, Data Governance council roles, and roles such as Data Product owners, Data Product developers, data platform representatives, and data privacy and security SMEs • Focuses on Metadata Management, Data Privacy, Data Protection, Data Security, Data Quality, Data Lineage, and Master Data Management • Leverages Machine Learning, AI and Knowledge Graph technologies, mature access control methods	• Supports a federated organization • Cross-domain policies are defined at a global level and executed locally by business domains • The majority of policies at the local level are defined and executed by business domains • Supports roles such as Data Product owners, Data Product developers, data platform representatives, and data privacy and security SMEs • Metadata Management, Data Quality, Data Privacy, Data Protection, and Data Security are a business domain's concern • Computational data policies are embedded into the Data Product's life cycle

Figure 3.1 – Data Fabric and Data Mesh comparison

We have evaluated the similarities and differences between Data Fabric and Data Mesh design. Let's explore how these two data management frameworks can be used in a complementary manner.

Data Fabric and Data Mesh's friendship

Data Fabric focuses on Self-Service data access via active metadata leveraging a composable set of tools and technologies. It offers the ability to discover, understand, and access data across hybrid and multi-cloud data landscapes with automation and Data Governance. It is primarily process and technology centric with flexibility in supporting diverse organizational models. On the other hand, Data Mesh is organizationally and process driven. It requires a technical implementation approach to execute its design. Data Mesh is at a higher level and Data Fabric is at a lower level. Data Fabric is capable of fulfilling Data Mesh's key principles.

Figure 3.2 provides a conceptual view where both the Data Fabric and Data Mesh designs are applied together in a Self-Service data platform.

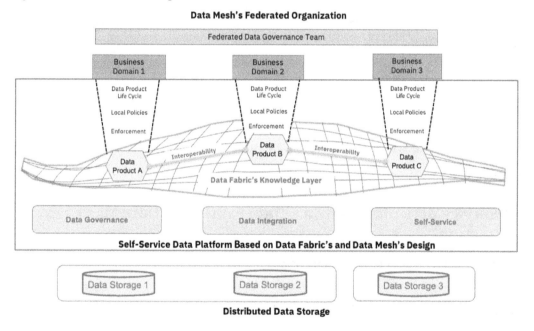

Figure 3.2 – Data Fabric with a Data Mesh Self-Service data platform

In the next four sections, I'll provide a point of view in ways a Data Fabric data architecture can apply Data Mesh's key principles. I'll discuss how to evolve this point of view into a technical architecture in *Chapter 9, Realizing a Data Fabric Technical Architecture.*

How Data Fabric supports a federated-based organization

Data Fabric is driven by active metadata and requires a business culture that prioritizes a metadata initiative, which involves capturing and managing metadata on an ongoing basis:

- Data Mesh's federated business domains are accountable for managing a Data Product's life cycle from development to end of life. The business domains can also take on the accountability to executing an active metadata driven culture in support of Data Products.

- Create a connected ecosystem of Data Mesh-based Data Products by domain with cross-domain relationships leveraging Data Fabric's knowledge layer

Let's look at how Data Fabric can satisfy Data Mesh's manage data as a product principle.

How Data Fabric manages data as a product

Data Fabric supports the concept of managing data as an asset, and managing data as a product. It achieves this with Data Integration, Data Governance, and Self-Service capabilities that manage the entire life cycle of data.

- Leverage Data Fabric's Data Governance building block to achieve a data consumer's understanding, trust, and use of Data Mesh's Data Products

- Apply Data Fabric's Data Integration and Self-Service building blocks to manage Data Mesh's Data Product life cycle while applying best practices in DataOps (automated testing/monitoring) and data as code (standards, policies, and infrastructure)

Let's look at how Data Fabric supports Data Mesh's Self-Service data platform principle.

Self-Service data platform via a Data Fabric and Data Mesh architecture

Data Fabric supports the autonomy required by Data Mesh business domains in a Self-Service data platform:

- Leverage Data Fabric's capabilities across the Data Governance, Data Integration, and Self-Service building blocks to enable scaling, sharing, and access to Data Mesh's Data Products

- Apply Data Fabric's federated Data Governance capabilities to isolate data management and governance in support of Data Mesh's business domains

Let's look at how Data Fabric can satisfy Data Mesh's federated computational governance principle.

Federated computational governance with Data Fabric

Data Fabric supports Data Mesh's federated computational governance principle by delivering value-driven Data Products that are secure and trusted:

- Data Fabric's Data Governance building block delivers the required automation, data policy management, and enforcement required in a federated organization

- Data Fabric's Data Governance building block addresses Data Mesh's Data Product usability attributes (discoverable, addressable, understandable, truthful, natively accessible, interoperable, valuable, and secure)

Let's summarize what we've learned in this chapter.

Summary

Data Fabric and Data Mesh are two separate but harmonious data architecture approaches. Both approaches represent sophisticated designs that speed up data delivery while ensuring data can be trusted and relied upon with the goal of extracting business value.

In this chapter, we highlighted how Data Mesh takes an organizational approach to managing data while Data Fabric takes a technical, metadata-driven approach. We identified how Data Fabric and Data Mesh focus on addressing data silos, data access bottlenecks, quality, security, and scalability issues. Both recognize the criticality of having strong Data Governance with embedded automation in the life cycle management of data. Data Fabric can be used together with Data Mesh to realize its core principles (domain ownership, data as a product, Self-Service data platform, and federated computational governance). We discussed how one approach does not need to be selected over the other. Both Data Fabric and Data Mesh can be used in a complementary manner to achieve rapid data access, high-quality data, and Data Governance.

In the next chapter, *Introducing DataOps*, we will provide an introduction to DataOps. DataOps is core to Data Fabric's design. It is a value-driven data management framework that further accelerates the life cycle management of data.

4
Introducing DataOps

DataOps is a collaborative data management framework that can be applied to a Data Fabric architecture by managing the stages from the creation to the deployment of data. It offers an operational model for data to deliver quality analytics. DataOps and Data Fabric can be executed independently of one another, but when used together, they create a high-quality, automated, cost-effective, and well-governed data environment on steroids. This chapter will introduce the DataOps framework and its principles. It will also look at its evolution from other popular frameworks such as DevOps, **Statistical Process Control** (**SPC**), and Agile. We will discuss the applicability of Data Quality to data observability, a subcomponent of DataOps, and the alignment of DataOps and Data Fabric.

By the end of this chapter, you will understand the basics of DataOps and how it aligns with Data Fabric.

In this chapter, we will cover the following topics:

- What is DataOps?
- DataOps' value
- Data Fabric with DataOps

What is DataOps?

DataOps is a framework that applies best practices, processes, and technologies using a collaborative approach to achieve fast, high-quality, and cost-efficient data delivery. Similar to Data Fabric, DataOps is not a tool or specific technology. Rather, it's a set of principles focused on managing data as code, which in turn enables a deep level of automation necessary for scaling out data management. It emphasizes teamwork across a diverse set of data roles working on data analytics to eliminate misalignment between teams. Its bedrock is the ongoing quality monitoring of both data and processes to achieve customer satisfaction and efficiency. It advocates reusability, iterative short deployment cycles, and feedback loops to achieve business and customer excellence.

DataOps is applicable to raw and business-ready data. It streamlines the development, testing, deployment, and monitoring of data and its pipelines by applying proven, successful quality control and software development principles.

DataOps' principles accelerate the business value derived from data analytics. Let's take a look at each of its principles.

DataOps' principles

The efficient and effective management of data has evolved over the past four decades. Many lessons learned have been incorporated into DataOps and other approaches. However, there is a standard set of DataOps principles established that provides a baseline. The **DataOps Manifesto** (https://dataopsmanifesto.org/en/) was established by *Christopher Berg*, *Gil Bengiat*, and *Eran Strod* with the objective of providing guidance to achieve successful, high-quality, and rapid data analytics focused on business value.

The following is a list of principles mentioned in the DataOps Manifesto:

- Continually satisfy your customer
- Value working analytics
- Embrace change
- It's a team sport
- Daily interactions
- Self-organize
- Reduce heroism
- Reflect
- Analytics is code
- Orchestrate
- Make it reproducible
- Disposable environments
- Simplicity
- Analytics is manufacturing
- Quality is paramount
- Monitor quality and performance
- Reuse
- Improve cycle times

The DataOps principles were inspired by other popular frameworks, such as Agile, DevOps, and SPC. In the next section, we will touch on each of these frameworks to provide an understanding on the evolution of DataOps.

The evolution of DataOps

DataOps' principles are inspired by three successful frameworks:

- DevOps
- Agile
- SPC

DevOps and Agile are successful software delivery frameworks. DataOps builds on the best practices of these two frameworks and spins its focus on the successful delivery of data and information. DevOps is a way of working, driving continuous improvement, bringing people together, breaking down barriers, and leveraging automation. DataOps is focused on the data aspects of these areas. One area of attention in DataOps is the need for Data Quality to be pristine where a quality control framework such as SPC applies. Let's review these frameworks in more detail.

Agile

Agile promotes communication, collaboration, and interaction across business and technical teams. It takes a self-organized team approach to manage workloads based on capacity. Software products are deployed frequently with quality to achieve high customer satisfaction. After each software delivery iteration, a self-reflection assessment is completed where the team identifies areas of success and improvements to deliver faster with higher quality in the next iteration.

Agile has 12 principles defined as part of the **Agile Manifesto** (`https://agilemanifesto.org/principles.html`):

- Our highest priority is to satisfy the customer through early and continuous delivery of valuable software.
- Welcome changing requirements, even late in development. Agile processes harness change for the customer's competitive advantage.
- Deliver working software frequently, from a couple of weeks to a couple of months, with a preference to the shorter timescale.
- Business people and developers must work together daily throughout the project.
- Build projects around motivated individuals. Give them the environment and support they need and trust them to get the job done.
- The most efficient and effective method of conveying information to and within a development team is face-to-face conversation.

- Working software is the primary measure of progress.

- Agile processes promote sustainable development. The sponsors, developers, and users should be able to maintain a constant pace indefinitely.

- Continuous attention to technical excellence and good design enhances agility.

- Simplicity--the art of maximizing the amount of work not done--is essential.

- The best architectures, requirements, and designs emerge from self-organizing teams.

- At regular intervals, the team reflects on how to become more effective, then tunes and adjusts its behavior accordingly.

Let's review the next framework, SPC.

Statistical process control

SPC is a proactive method that leverages statistics to measure and monitor process quality to drive performance improvements. It is tied to **Statistical Quality Control** (**SQC**). In his book *Information Quality Applied*, Larry English defines these methods as follows:

SPC is "*the application of statistical methods, including Control Charts, to measure, monitor, and analyze a process or its output in order to improve stability and capability of the process, a major component of Statistical Quality Control.*"

SQC is "*the application of statistics and statistical methods to assure quality, using processes and methods to measure process performance, identify unacceptable variance, and apply preventive actions to maintain process control that meets all Customer expectations...*"

SPC has been heavily applied in the Japanese car manufacturing industry. In the 1950s, Japanese car manufacturers struggled to make a profit and drove a culture change to adopt quality improvement principles that achieved efficiency with higher quality, now called **Lean manufacturing**. Lean manufacturing was successfully adopted and credited for Toyota's success; it is now one of the largest car manufacturers in the world.

These same quality principles that worked in the manufacturing of cars also work in the manufacturing of data, which is why they are an integral part of DataOps. The objectives focus on the proactive measurement and monitoring of the quality of data in real time to drive process and Data Quality improvements. Applying SPC and SQC best practices is not new in DataOps; they were advocated by Larry English in Data Quality management within the **Total Information Quality Management** (**TIQM**) system, *Process 2: Assess Information Quality*. The premise is to assess the quality of raw and business-ready data and then identify and address the root cause with process improvements, as opposed to correcting and cleansing data only as a one-time tactical effort.

DevOps

DevOps combines **Development** (**Dev**) and **Operations** (**Ops**) to speed up the release of software products. Similar to Agile, it involves a culture change that advocates for development and operational teams to collaborate and derive greater value. The objectives are to develop higher-quality solutions with stable and quality deployments at a lower cost faster. DevOps' core components include iterative deployments that apply **Continuous Integration** (**CI**), **Continuous Delivery** (**CD**), and **Infrastructure as Code** (**IaC**). DevOps borrows best practices from Agile and SPC as well as Lean manufacturing. The objectives are to replace traditional waterfall approaches in software delivery with short and frequent deployments that provide value more quickly. DevOps is an iterative and continuous framework for developing, testing, deploying, and monitoring code.

I have referenced some of the key DevOps principles highlighted by authors *Noah Gift* and *Alfredo Deza* in their book *Practical MLOps*, where they discuss the applicability of DevOps in MLOps, which can also be applied to DataOps:

- **Continuous integration**: Automated continuous testing of a software solution
- **Continuous delivery**: Automated continuous deployment of a software solution
- **Microservices**: An independently deployable service that isn't coupled to other services
- **Infrastructure as code**: A programmatic approach to managing hardware such as compute, servers, and storage
- **Monitoring and instrumentation**: Monitors the quality of a software solution

Now that we know the background of the three frameworks that gave rise to DataOps, let's review the people, process, and technology dimensions that compose a DataOps framework.

DataOps' dimensions

DataOps focuses on agility and efficiency by identifying non-value-added areas in a data life cycle. The goal is to drive improvements across people, processes, and technology. It identifies better ways of working, collaborating, and communicating – the *people dimension*. It leverages proven best practices based on principles that generate rapid, high-quality solutions while reducing cost – the *process dimension*. It ensures the tools and technologies have the right capabilities to deliver DataOps-based solutions – the *technology dimension*.

I'll provide a perspective that takes the DataOps principles and slices them into the applicable people, process, and technology dimensions. The reality is that these principles are not always clean-cut and aligned to a particular dimension and in some cases may apply to all three. However, I have organized them based on the closest alignment of a dimension to a principle. *Harvinder Atwal*, in his book *Practical DataOps*, provides a detailed explanation where he classifies DataOps principles into *Ways of Working*, *Technical Practices*, and *Technology Requirements*. It provides a good base for thinking about these principles in terms of applicability. I'll talk about some of those thoughts in the point of view I offer.

In *Figure 4.1*, we have organized each DataOps principle into the dimension it has the strongest alignment with:

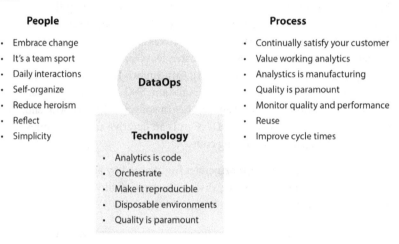

People

- Embrace change
- It's a team sport
- Daily interactions
- Self-organize
- Reduce heroism
- Reflect
- Simplicity

DataOps

Technology

- Analytics is code
- Orchestrate
- Make it reproducible
- Disposable environments
- Quality is paramount

Process

- Continually satisfy your customer
- Value working analytics
- Analytics is manufacturing
- Quality is paramount
- Monitor quality and performance
- Reuse
- Improve cycle times

Figure 4.1 – DataOps principles organized by people, process, and technology

Let's dive deeper into each of these dimensions, starting with people.

People

This dimension enables the effective collaboration and communication of all data roles in the development and deployment of data analytics. There is an emphasis on culture change to derive value from each data role to more effectively deliver data. There is a need for better cross-team alignment among the diverse data roles working on building data and data pipelines. To scale and iterate quickly, eliminating bottlenecks and quickly adapting to evolving business requirements are necessary. Teams work better and more effectively by independently executing work as needed and fine-tuning execution performance based on customer feedback.

The roles typically involved in DataOps are as follows:

- **Data producers (data engineers, DataOps engineers, and architects)**: DataOps engineer is a role that has evolved from both the software engineer and data engineer roles. They are **Subject Matter Experts** (**SMEs**) in DataOps best practices and principles and work with data engineers, data analysts, and data scientists to deploy data pipelines that produce Data Assets and Data Products.

- **Data consumers (business or data analysts and data scientists)**: This role is necessary to deploy a successful analytics solution. A data consumer can also become a data producer if they consume data, and compose new data that can then be shared.

In addition to these roles, support from executive business leaders to drive change is required to achieve success in a DataOps discipline.

The following are the DataOps principles aligned with the people dimension:

- **Embrace change**: Leveraging the agile methodology, this principle welcomes evolving customer requirements and direct communication with the customer.

- **It's a team sport**: Embrace the diversity of data roles, experiences, and skills to build innovative and productive data analytics. This also includes the diversity of tools preferred by each role.

- **Daily interactions**: Daily communication across customers, analytics teams, and operations is necessary to keep a pulse on what's happening, understand the needs of all parties, and drive necessary actions.

- **Self-organize**: Teams are trusted to independently manage their own work based on their capacity.

- **Reduce heroism**: Data roles work together using the right set of tools and practices with the objective of scaling quickly. Eliminate bottlenecks where one person or team takes on more than they can handle; agility is key. Remove wasted time spent firefighting issues.

- **Reflect**: Review lessons learned, customer feedback, and successes and implement steps to drive further improvements to increase productivity.

- **Simplicity**: Places an emphasis on creating simple but effective designs. Use technologies where needed without creating complexity.

Next, let's review the process dimension.

Process

DataOps processes are focused on high speed, iterative deployments, short cycle times, and quality. The process dimension expedites and improves the development, testing, deployment, and monitoring of data solutions to deliver high value. Getting customer feedback early across iterative short cycles offers the opportunity to refine the product as needed to better meet customer needs.

The following are the DataOps principles aligned with the process dimension:

- **Continually satisfy your customers**: Deliver data insights in smaller chunks iteratively to get customer feedback and validation.

- **Value working analytics**: Insights delivered need to be accurate and provide value to the business. Data trust and reliability are key.

- **Analytics is manufacturing**: Leverage quality control best practices from Lean manufacturing and apply them to data analytics. Quality control achieves efficiency and quality.

- **Monitor quality and performance**: Quality, security, and performance require ongoing proactive monitoring and immediate alerting to drive action and resolution.

- **Reuse**: Apply reuse wherever possible in analytical and process design to deliver quicker. Avoid re-inventing the wheel and replace manual activities with highly automated processes.

- **Improve cycle times**: Aim to develop and deploy analytics in short cycle times to achieve high delivery speed.

Let's take a look at the third dimension, technology.

Technology

Technology is an enabler of the people and process dimensions. It provides the means of achieving DataOps by ensuring the right features and capabilities are available through the right set of tools and technologies. However, technology is one dimension of the three and you need all three to be successful. The bottom line is, from a technology standpoint, there is a need for a diverse set of tools and technologies, whether private or open source. The goal is to manage data seamlessly from end to end, with speed, in an automated way with high quality and governance.

The following are the DataOps principles aligned with the technology dimension:

- **Analytics is code**: The use of code-based tools with version control and IaC offers a high level of automation, provides an audit trail for changes, and enables a modular design such as the use of parameters. It offers the use of CI/CD, which enables reproducibility, refactoring, and collaboration.

- **Orchestrate**: Allows the end-to-end orchestration of data pipeline activities across a variety of data types, data formats, tools, technologies, and data environments.

- **Make it reproducible**: Reproduce all activities across data, code, and environments using version control and CI/CD to minimize risk during production deployments.

- **Disposable environments**: Provides sandbox environments that can be easily created and disposed of without relying on central support teams to enable rapid innovation.

- **Quality is paramount**: Automated Data Governance (Data and Metadata Quality, Data Lineage, Data Privacy, Data Protection and Data Security, Master Data Management) that monitors in-flight data and identifies information, security, and configuration issues.

In the next section, we'll describe two frameworks that depend on DataOps-based data pipelines: MLOps and AIOps.

MLOps and AIOps depend on DataOps

The three frameworks – Agile, SPC, and DevOps – that gave rise to DataOps also inspired additional xOps types of frameworks. In some cases, they are further classifications of DataOps, such as Data Governance, called **DataGovOps**, or Data Security, called **DataSecOps**. MLOps and AIOps are two frameworks that depend on prepared high-quality and well-governed data, and thereby the outputs produced by DataOps pipelines. Let's review what each one focuses on.

MLOps

MLOps refers to the development, testing, deployment, and monitoring of **Machine Learning** (**ML**) solutions. It includes a series of steps in the life cycle of data where it intersects with an ML solution. DataOps and MLOps integrate with the preparation of data. Without data, it is an empty shell of a non-working ML solution. DataOps enables the necessary data preparation and delivery to enable feature engineering and model testing in MLOps. MLOps leverages principles such as model version control and CI/CD. It focuses on the automated testing of results for accuracy. High-quality principles in DataOps are what MLOps relies on. It requires and expects data to be trustworthy to provide high confidence levels and quality predictions, otherwise the ML solutions fail. Data observability is applicable in MLOps due to the dependency on high-quality data. Some of the areas MLOps monitors are Data Quality, data drift, and data bias.

AIOps

AIOps is often confused with MLOps, although the distinction here is that AIOps involves the automation of service desk operations leveraging technologies such as ML and big data. Some of these tasks might include understanding and analyzing system and application defects in an automated manner to identify correlations and drive resolution. AIOps has dependencies on both DataOps and MLOps. The common denominator is high-quality, business-ready data produced by DataOps.

We now have a general overview of DataOps. In the next section, we'll focus on the business value it derives.

DataOps' value

DataOps accelerates operations in building and delivering data solutions. It can be viewed as a layer sitting on a Data Fabric architecture that expedites data processing and delivery to achieve an organization's digital transformation journey. As discussed in *Chapter 2*, *Show Me the Business Value*, there are four key ingredients to achieve profitable data monetization:

- Trusted quality data
- Meaningful insights
- Action-oriented business plan
- High execution speed

DataOps with Data Fabric focuses on the delivery of trusted, quality data and meaningful insights, that is, insights that can be acted upon to derive value and reliability. Both the first and second points enable the creation of a lucrative business plan that can be executed on. Time is money, and DataOps and Data Fabric specialize in achieving the preceding four goals quickly.

The following is a summary of the key objectives of a DataOps discipline:

- Customer satisfaction
- Agile data-driven culture
- Fast delivery of data insights
- High Data Quality
- Few system errors
- Low cost
- High performance and productivity
- Efficient collaboration across data practitioners in business and IT
- Trusted and reliable business decisions

DataOps focuses on improving overall Data Quality. **Data observability**, a subcomponent of DataOps, entails the execution of proactive quality monitoring with the intention of driving immediate resolution. Data Quality is the foundation of data observability. However, data observability takes a modern approach to ensuring Data Quality as opposed to a traditional approach.

Let's understand the distinction between traditional and modern Data Quality.

From traditional Data Quality to data observability

Data Quality has a decades-long, rich history. It was recognized that improving overall business productivity starts with improving processes and Data Quality. Larry English's TIQM framework is based on this premise. TIQM was first introduced in 1999, with English's first book titled *Improving Data Warehouse and Business Information Quality*.

Larry English's TIQM is centered around two key objectives:

- Quality analysis of metadata, data, and processes to identify the instances of issues
- Drive course correction in the form of removing the instances of errors and, most importantly, identifying the root cause to address it with process improvement

He describes TIQM objectives as follows:

> *TIQM is a complete, yet practical system of quality principles, processes and techniques that can be easily applied to measure Information Quality and improve processes to eliminate the causes of poor quality information and its resulting process failure, losses, and costs.*

What's notable about TIQM is that even though it was introduced three decades ago, these foundational concepts are applicable today. Let's dive into the six TIQM processes and understand how TIQM, a Data Quality framework, is relevant and should be leveraged in data observability.

TIQM's six Data Quality processes

TIQM has six processes, listed here, that focus on quality assessments for metadata, data, information, and processes. It provides a prescriptive approach to Data Quality assessments and remediations:

- Process 1: Assess information product specification quality

- Process 2: Assess information quality

- Process 3: Measure poor quality information costs and risks

- Process 4: Improve information process quality

- Process 5: Correct data in source and control redundancy

- Process 6: Establish the information quality environment

> **Note**
>
> TIQM draws a distinction between the terms "data" and "information," where data is raw data and information is business-ready data. Hence, there is a distinction between the terms "information quality" and "Data Quality". The same applies to English's concept of "information products." In this book, I use the terms data and information interchangeably.

TIQM focuses on quality assessments that drive data correction and process improvements that address the root cause of the data errors, as opposed to a temporary solution such as data cleansing, which Larry English calls it, a non-value-added cost of poor Data Quality. Every data correction needs to have an identified root cause of the issue with a process improvement effort in parallel that addresses it.

Process 1: Assess Information Product Specification Quality and *Process 2: Assess Information Quality* identify a number of important attributes focused on data/metadata trust, intuitive business understanding, and reliability.

Here are some attributes that require quality analysis for metadata in *Process 1* and data content in *Process 2*:

- Process 1 – Metadata quality:

  - Business term

  - Entity type

  - Attribute

  - Business rule

- Process 2 – Data content quality:

  - Accuracy

  - Completeness

  - Validity

  - Uniqueness

  - Timeliness

  - Data bias, called presentation objectivity

Many assumptions are made about traditional Data Quality when compared to modern Data Quality. Let's review what they are.

The distinction between traditional Data Quality and modern Data Quality

Traditional Data Quality assumes a centralized Data Governance organization. This assumption is based on a centralized data management approach and doesn't take into account distributed data architectures introduced predominately by cloud technologies. Traditional Data Quality assumes the use of relational data storage and relational data modeling, which in turn also assumes data ingestion and data preparation based on **Extract, Transform, and Load** (ETL) operations. It does not consider an **Extract, Load, and Transform** (ELT) approach.

Another consideration is Data Quality for streaming data, which is drastically different as it doesn't always provide a complete picture of the data and may require different progression levels of the data to achieve good Data Quality. The techniques and tools used to analyze, cleanse, and monitor Data Quality have drastically evolved. There is a deep need to automate, scale, and gauge the state of quality at the process, data, and metadata levels. Tools and technologies need to take into account the four Vs of data introduced in *Chapter 1* (volume, velocity, variety, and veracity) to keep up. Data Quality analysis based on a point-in-time snapshot of data is not sustainable and won't provide a view of the health of data. There is a high demand to know the quality of data across a proliferation of sources and data types with huge data volumes.

Data observability takes a proactive approach by monitoring the quality of in-flight data and processes. It doesn't wait for the data to be captured and consolidated into data storage like a database, data warehouse, or data lake. It monitors the data as it flows through the pipelines from target to source as the transformations are applied to identify failures and issues such as data errors, poor performance, or security. It provides immediate alerts and notifications with sufficient information and details to enable immediate resolution. Data observability provides an execution approach but doesn't provide a prescriptive approach on the criteria of how and what to analyze regarding Data Quality. This is where Data Quality frameworks such as TIQM can be leveraged together with data observability to have complete coverage in executing Data Quality at rest and in flight.

While not exhaustive, *Figure 4.2* captures key distinctions between traditional Data Quality and modern Data Quality:

Traditional Data Quality	Modern Data Quality
<ul><li>Centralized data management</li><li>Top-down Data Governance model</li><li>Relational data storage with an enterprise model</li><li>Relational data modeling with inherent relationships</li><li>ETL only</li><li>Data Quality analysis based on a point-in-time static data snapshot</li><li>Manual and lengthy defect resolution</li></ul>	<ul><li>Supports both centralized and decentralized data management</li><li>Central, federated, or distributed Data Governance model</li><li>SQL or NoSQL data storage (data warehouses, data lakes, and data lake houses)</li><li>Relational, **Resource Description Format (RDF)** (and inferenced relationships), and graph data modeling</li><li>ETL, ELT, streaming, and replication</li></ul>

Figure 4.2 – Traditional Data Quality compared to modern Data Quality

Now that we have a comprehensive understanding of the relationships between data observability and Data Quality, let's tie this back to Data Fabric to understand how DataOps aligns to derive greater value.

Data Fabric with DataOps

To support one of the key principles of a Data Fabric architecture, *data are assets that evolve into Data Products*, Data Fabric needs to be married with DataOps principles that focus on quality control and efficiencies in the development and delivery of data. DataOps applies principles in how data and data pipelines are developed, such as with embedded automation and orchestration. It expects the testing of data, assumptions, and integrations at a grand scale. It establishes quality controls from initial data transport from the source to the destination(s), and then after that, executes ongoing data monitoring. Data is fluid; the business evolves and requirements change, which requires proactive monitoring and analysis to ensure the successful delivery of data to consumers.

The DataOps phases are iterative and are not always executed in sequence. Sometimes, a phase may be executed multiple times. For example, during testing, you might find data validation defects that require updating the transformation code as part of the develop phase. The same applies to the testing of the orchestration flow that may require refactoring and configuration changes.

Figure 4.3 is a conceptual diagram that portrays data throughout the DataOps phases.

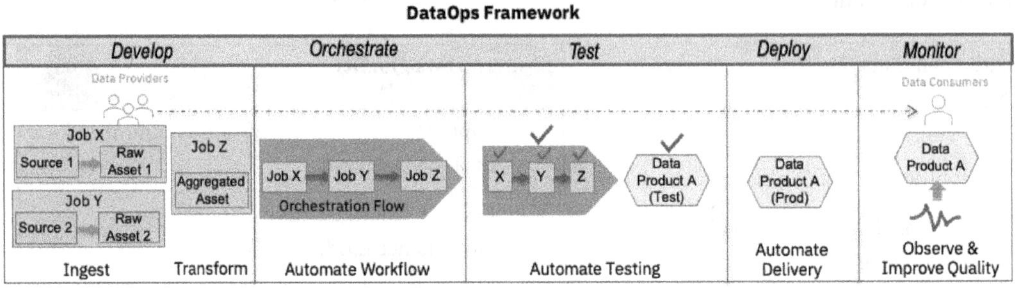

Figure 4.3 – Development and Deployment of Data via DataOps

The DataOps *people*, *process*, and *technology* principles apply to every stage of data as it travels through a Data Fabric architecture from creation to end of life.

Let's review each DataOps phase and how Data Fabric plays a role.

Develop

In the develop phase, a Data Fabric's *Data Governance* building block focuses on ensuring high-quality, secure, and trusted data is ingested, prepared, and delivered. It also offers the knowledge to make decisions on integration approaches based on data and metadata relationships. Data Fabric's *Data Integration* and *Self-Service* building blocks enable the means to facilitate data processing and delivery activities.

Orchestrate

The orchestrate phase takes the jobs, data pipelines, and dependent components such as infrastructure and automates a workflow. This entails automating the execution of jobs, following a specific sequence, and can include conditional statements based on the output or status of jobs. It also includes setting up infrastructure or other dependent processing components. It facilitates the means for data to be managed from end to end across different data environments, tools, and technologies. The orchestrate phase intersects a Data Fabric's *Data Governance* and *Data Integration* building blocks in that data needs to be governed across disparate sources/pipelines and technical approaches.

Test

Automated testing and quality control are key best practices in this phase. Data, metadata, configurations, integrations, and code for data pipelines, orchestration workflows, and IaC are tested in this phase. It ensures each job works independently, and as a whole with its relationships and dependencies. Based on the results of this phase, there can be several iterations between the *test*, *orchestrate*, and *develop* phases. All three Data Fabric building blocks are applicable here. Testing is carried out across data ingestion, data integration, and data consumption activities.

Deploy

The second-to-last phase in DataOps is the automated deployment of data. This involves the automated delivery of assets/data products (data pipelines, workflows, data, configurations, metadata, and so on). This is where all the work in the previous phases comes together and is presented as production-ready analytical data.

Monitor

The last DataOps phase is monitor. In traditional Data Quality practices, quality control postproduction is often neglected. The premise is once you promote something into production, you're done. That's not the reality. Data observability monitors data in production to ensure it's reliable and available with high quality. This needs to be executed across the entire Data Fabric architecture.

Let's quickly summarize what we've learned in this chapter.

Summary

Applying a Data Fabric architecture with a DataOps framework dramatically raises the bar to deliver data with high agility, quality, and governance at a lower cost. In this chapter, we provided an introduction to DataOps, its value, and the 18 driving principles. We briefly introduced Agile, SPC, and DevOps from which DataOps borrows many of its principles. We reviewed the distinction between traditional Data Quality and modern Data Quality, and how modern Data Quality can leverage a foundational Data Quality framework. We also defined data observability and its relationship to Data Quality. We concluded by conceptualizing the use of a Data Fabric architecture together with a DataOps framework. Both offer tremendous value and should be applied in the life cycle management of data.

At this point in the book, we have introduced Data Fabric, Data Mesh, DataOps, and foundational Data Governance concepts. In the next chapter, we will focus on a key business artifact, a data strategy that every organization should have as part of their digital transformation journey.

5

Building a Data Strategy

A data strategy articulates an organization's plan for reaching high data maturity. Data maturity is the measurement of an organization's expertise in applying data capabilities to achieve business value. It first requires establishing a data maturity baseline and identifying a target data maturity level. Progress needs to be measured frequently with course correction as needed. Continuous and iterative progress are the objectives of a data strategy. Achieving success with a data strategy requires a focus on data management, Data Governance, Data Privacy, Data Protection, Data Security, and other critical aspects for an organization to succeed in its digital transformation journey. There needs to be careful thought on how to execute and implement a data strategy.

In this chapter, we'll guide you in creating a data strategy that is applicable across industries. It offers tips to create, execute, and implement a data strategy. It also provides guidance on the role of a Data Fabric in a data strategy. By the end of this chapter, you will have the basic tools to create a data strategy document and position a Data Fabric solution.

In this chapter, we'll cover the following topics:

- Why create a data strategy?
- Creating a data strategy
- Data strategy implementation

Why create a data strategy?

There are many reasons why an organization needs a data strategy. The most obvious one is to have a plan on how to become data-driven in order to generate revenue and achieve cost savings. Enterprises have learned from the pitfalls of not managing data as an asset and data product. They have lost customer opportunities and revenue from business decisions made with untrustworthy data or experienced high costs when developing customer goods. In *Chapter 2, Show Me the Business Value*, we discussed examples such as costs associated with regulatory fines due to lack of Data Privacy enforcement or lost revenue from bad metadata and Data Quality. These factors lead us to the need for a well-thought-out and structured data strategy that manages data effectively and is aligned with the enterprise's business goals.

Not having a data strategy document can lead to the following negative outcomes:

- An immature level of organizational readiness
- Incorrect business decisions based on poor quality data
- A tactical and reactive approach to digital transformation leading to failures from a lack of vision
- Data silos and IT bottlenecks
- Non-compliance with regulations
- Lack of Data Governance creating risks to data monetization

The best way to understand why a data strategy is necessary is by first recognizing the value of managing data as an asset and a data product. It's all about the dollar signs: *profit*. Data insights that can be leveraged to make strategic business decisions and that result in revenue are the goal. I'll leave you with a quote from *Ian Wallis* from his *Data Strategy* book where he uses a water analogy to describe the value of data:

> *"...think about the value of water and the parallels with data increase. Champagne is 85 percent water ... whether it is the UK's favourite Moët & Chandon at £38 a bottle, or a Taste of Diamonds 2013 at £1.2 million – consider how much value has been added to that 85 per cent water content. Then think about the parallels with data, how data captured at minimal cost in the course of an interaction with a supplier, customer or other party can support decisions worth millions to your organisation."*

In the following section, we will review how to assess an organization's data maturity level.

A data maturity framework

A prerequisite to defining a data strategy is to understand the current state of an organization's data maturity based on its goals, initiatives, standards, and capabilities. **Data maturity** is defined by an organization's level of expertise in data management practices, data skills, and data capabilities to derive insights that achieve business value. A data maturity assessment provides a baseline measurement of where an organization is on its data journey. It is an indicator of the data management health of an organization.

> **Note**
>
> A maturity model is a great tool for acknowledging the current maturity state of an organization and measuring improvement. However, it's important to understand that it shouldn't be used to achieve an end state. There is always room for improvement, and continuously measuring progress and improvement should be the objective.

A maturity assessment should review all aspects of data management taking a people, process, and technology approach. It should target a diverse set of organizational functions for better representation.

Some questions to consider when conducting a data maturity assessment are as follows:

- People:

  - Are there clear roles with designated responsibilities in place to achieve these tasks?

  - Do these roles have the necessary skill sets and experience?

  - Is there a clear understanding between the business and technical teams on what's required to achieve these tasks?

- Process:

  - To what degree are frameworks, principles, standards, and best practices applied and leveraged to complete these tasks?

  - Is there a Data Governance program in place to execute a checks and balances system to execute these tasks? Are various operational models supported?

  - To what degree are business processes data-driven? Do business and data processes work together in a cohesive manner?

- Technology:

  - Are there strategic tools and capabilities in place to target key focus areas such as Data Quality, Data Lineage, Metadata Management, Master Data Management, Data Privacy, Data Protection, and Data Security?

  - To what degree are Data Governance and Data Integration tasks automated?

  - Is there a well-thought-out strategic data architecture that focuses on the automation of Data Management and Data Governance?

There are a number of standard and common data maturity models available. One is the **Data Management Body Of Knowledge (DMBOK)** by **Data Management Association (DAMA)** International (`https://www.dama.org/cpages/body-of-knowledge`), another is the **Data Management Capability Assessment Model (DCAM)** from the EDM Council (`https://edmcouncil.org/frameworks/dcam/`), and there is also Gartner's **Enterprise Information Management Maturity Model** (you may need to subscribe to access this link: `https://www.gartner.com/document/3236418?ref=hp-wylo`). Each has a different breadth of scope in the areas it covers and slight variations in how it describes each maturity level.

An organization can select a framework that best aligns with its business objectives and desired data capabilities. I would encourage you to adopt an existing one, and if needed, tweak it to better align with your needs. Why reinvent the wheel? There has been an extensive amount of work put into many of these frameworks. A tip is to ensure the framework has been continuously updated to reflect the changing market. Typically, data maturity frameworks have a five-level rating system of maturity for each area of focus. Here is a point of reference that compares many of the top industry data maturity models: `https://datacrossroads.nl/2018/12/16/data-management-maturity-models-a-comparative-analysis/`.

For perspective, *Figure 5.1* displays DAMA's DMBOK maturity model, which is data management focused and has 12 areas, the DCAM, with 6 areas, and Gartner's, which is at a higher level and has 7 areas:

DMBOK	DCAM	Gartner
• Data Governance • Data Quality • Data Architecture • Data Modeling and Design • Data Storage and Operations • Data Security • Data Integration and Interoperability • Document and Content Management • Reference and Master Data • Data Warehousing and Business Intelligence • Metadata	• Data Strategy and Business Case • Data Management Program and Funding • Business and Data Architecture • Data and Technology Architecture • Data Quality Management • Data Governance	• Vision • Strategy • Metrics • Governance • Organization and Roles • Lifecycle • Infrastructure

Figure 5.1 – Data maturity model focus areas for the DAMA, the EDM Council, and Gartner

In the next section, we will discuss what is necessary to get started with a data maturity assessment.

A data maturity assessment

A data maturity level represents an organization's current experience in managing and using data. Using a map and distance analogy, an organization needs to know what its starting point is – the current data maturity level – to determine a destination point. This is how navigation systems work today; you enter a starting point and a destination and then you have a list of directions, which, in this case, would translate to a data strategy. An organization might perceive its maturity level to be a lot higher than it truly is. Having a Data Governance or data analytics program doesn't always mean high data maturity. What matters is how you do it and whether it produces business value. A data maturity assessment gives you an understanding of the gaps and what's working and not working and serves as a starting point. It also helps to establish a target data maturity level.

To get started with a data maturity assessment, there are a few things that need to be decided:

- Determine the participants – which business units, departments, or individuals do you seek input from? Prioritize participants based on the level of impact the issues and challenges in their areas cause. This can be prioritized by the areas that exhibit the most issues whether poor customer satisfaction, low profit, or high production costs.

- Is all feedback equally important? Determine the weight given to each participating business unit, department, or individual. Input might be more critical for one area than the other based on its day-to-day focus.

- Establish a threshold for each focus area – what is the minimum acceptable data maturity level for each focus area? This will help establish the ratings that are out of bounds, which, in turn, become candidates for improvement.

A data maturity assessment needs to be sincere and provide an accurate reflection of the as-is state of data management. It needs to capture the good with the bad to determine the current gaps. The focus of this exercise is not to identify blame but rather to identify potential non-value-added areas, define improvement, and create an approach as part of a data strategy. If you recall, the same mindset is applied to the assessment of information quality. This was discussed in *Chapter 4, Introducing DataOps*, as part of the **Total Information Quality Management** (**TIQM**) framework. A data maturity assessment will determine how well existing initiatives function with respect to data management. It will provide a measure across areas of focus based on the selected data management framework.

In the next section, we will discuss an approach to creating a data strategy.

Creating a data strategy

A data strategy document's focus is driven by an organization's business goals, current as-is data maturity level, aspired data maturity, and associated capabilities. As you have seen in this book, the common theme is always Data Governance to achieve data integrity, which is a required topic in a data strategy document. The other is overall data management. Data management can be broken up into different compartments such as Data Integration, data operations, data architecture, and so on. Regardless of which focus areas are added, what's critical is to ensure it always ties back to the business. It needs to address how to achieve the desired level of data maturity and how the associated capabilities will accelerate an enterprise's business objectives and vision. One common pitfall that I have seen over and over when working with brilliant technical practitioners is the inability to tie technology back to business value. It's not about how cool technology is; it's about how that innovative technology creates business value and leads to data monetization.

In the next section, we will review what topics should be addressed in a data strategy document.

Topics in a data strategy document

The topics to include in a data strategy document can vary by the organization based on its aspirations and goals. As mentioned earlier, all data strategy documents need to address data management and Data Governance. The following are a set of topics suggested by the author *Ian Wallis* (in *Chapter 2* of his *Data Strategy* book) that can be used as reference:

- An executive summary
- The data management vision
- The business case
- Alignment with the corporate vision or overall business drivers
- The business goals realized by data
- Evaluation criteria to track progress
- The data strategy scope to be delivered
- Roles and responsibilities
- Risks, assumptions, and dependencies
- Data management:

  - Data standards
  - Data architecture
  - Data Governance and quality
  - Data Integration and migration
  - Data acquisition
  - Data transformation and exchange
  - Data compliance
  - Data accessibility
  - Master Data Management
  - Metadata Management

- Data exploitation:

  - Reporting/**key performance indicators** (**KPIs**)/dashboards
  - Analytics/data science/machine learning/artificial intelligence
  - Data insights
  - Knowledge management

I added metadata management to the sublist of data management, as it must have a dedicated focus. How you address metadata quality compared to data content quality involves different approaches that need to be articulated in a data strategy document, and both need to have a dedicated focus. The total number of pages for a data strategy document can also vary. Making the document too long will cause people to gloss over it and make it less likely to be read and consumed. If you make the document too short, it won't provide the value it was intended to achieve. Wallis suggests anywhere from 12-20 pages, which provides a good balance of information, not too short and not too long.

In the next section, we will discuss the process of creating a data strategy document.

Creating a data strategy document

Executing the creation of a data strategy document can be a daunting task. It requires a lot of coordination, perseverance, and negotiation with key stakeholders, sponsors, and SMEs. The creation of a data strategy will require a deep level of stakeholder involvement. There will be several iterations before the first final version of a data strategy document is complete. Remember, data strategy documents are living documents and need to continuously evolve with the business where new areas of improvement are identified. It also needs to be written and communicated in an easy-to-digest manner. The approach to creating a data strategy document should take a business point of view that connects data and its supporting technical components. This will make it more likely to get the necessary buy-in from key stakeholders.

The goal when working with stakeholders is to transition them from stakeholders into advocates for the data strategy. For advocates to talk about the data strategy, promote it and enable others to understand and embrace it. This will secure the success of the implementation of a data strategy. A constant feedback loop needs to always be welcomed, whether in the form of criticism or praise. There needs to be a balance with how the feedback is consumed – whether it provides value and generates change or just needs to be captured with no immediate action taken. Any transformation effort in an enterprise will always have some naysayers who haven't yet embarked on the necessary culture change and who will create resistance. On the other hand, when stakeholders raise valid points in the interest of achieving success, they need to be carefully considered and included in a data strategy.

The following are some tips to consider in the process of creating a data strategy document:

- **Business language**: Write in the context of the business value that data and the proposed capabilities will achieve.

- **Vision:** Clearly articulate the short- and long-term vision. Where do you want to be in a year? Five years? Over 10 years? A short-term vision needs to be more detailed and specific in terms of what that means, compared to a long-term plan providing room for change as the business evolves.

- **Stakeholder buy-in**: Recruit and engage the right stakeholders. Work to get their agreement on the value of creating a data strategy and convert them into advocates.

- **Sponsorship**: To succeed, you need the right sponsors to embrace and agree with the data strategy. Ideally, the sponsors will influence cultural change, and help tackle potential blockers that delay the progress of creating or implementing a data strategy.

- **Scope**: The scope of the data strategy document needs to be clearly articulated. What are the departments, processes, applications, systems, data, and other business aspects a data strategy needs to focus on? Areas of improvement should be identified for data management, Data Governance, Data Protection and Data Privacy. Existing pain points faced by the organization should be aligned with these focus areas, and the approach to how they will be addressed should be decided on and documented.

- **Project management**: Answer the who, what, and by when questions. Who will create a data strategy document? What are the tasks necessary to execute the data strategy? By when do these tasks need to be completed? The key is to ensure progress is tracked. The aim is not to lose sight of the goals and keep everyone progressing; otherwise, it can lead to the failure of a data strategy.

- **Data maturity framework**: Leverage a data maturity assessment framework to establish a baseline and continue progress. Use the data maturity assessment results as input to measure progress over time based on a target. This is a continuous activity, as there is always room for improvement.

- **Communication**: Ongoing communication is key. The status, progress, and next steps need to always be communicated. Communication keeps the momentum going to ensure stakeholders are on the same page. This should be in the form of verbal, written, or visual communication.

- **Living document**: The data strategy is a living document. It should never be considered static. As the business evolves, so does the data strategy document. It's important to iterate as needed and make the latest version of the document accessible.

- **Address the three dimensions**: The data strategy must take a people-, process-, and technology-based approach. It shouldn't be overly focused on one dimension, but rather all three to be successful.

- **Feedback and refine**: Similar to what we have learned about software development, DataOps, and other frameworks, the same applies to a data strategy. Feedback is bliss; lessons learned need to be documented. A data strategy needs to be refined as needed, including what has been learned about areas for improvement.

In the next section, we will focus on what to consider when implementing a data strategy.

Data strategy implementation

At this stage, a data strategy document has been created. The stakeholders are all on board and the sponsor(s) is rallying the implementation of the data strategy. The next step is agreeing on how to implement a data strategy. You can leverage the data maturity assessment results and target data maturity levels as a guide. These two parameters can be used to define new improvement initiatives and implementation approaches.

As an example, if the target state is to achieve a high level of data maturity in Data Quality and Data Integration to address company-wide data breaches, then these focus areas need to be aligned with potential solutions from people, process, and technical standpoints. Some of the questions to consider when selecting a solution are as follows:

- Can data practitioners collaborate better? Are the interactions needed between technical roles and the wider business addressed?

- What are the necessary process changes to drive faster production of customer goods with higher quality?

- What successful best practices and principles in data management can be applied across Data Governance, Self-Service, Data Privacy and Data Security?

- What data management tools elevate automation across Data Quality, Data Governance, Data Integration, and more?

- How and what supports managing data as an asset and as a data product?

A Data Fabric solution can play a key role in an organization's data strategy. In the first four chapters of this book, we introduced what a Data Fabric architecture is and its value statement, and we introduced other frameworks such as a Data Mesh and DataOps, which could be used together with a Data Fabric architecture. These are all examples of approaches and frameworks that could be positioned to address existing weak points surfaced by a data maturity assessment. The Data Fabric architecture offers tremendous value in managing data as a product and asset and instilling automation. It enables Data Protection, and Data Privacy, ensuring Data Security is in place and managing data across all the stages in its life cycle with embedded Data Governance and Self-Service.

Figure 5.2 illustrates a few of DAMA's data maturity areas mapped to the building blocks/architecture of a Data Fabric. This can be a first step when considering solutions and approaches. Start with a solution gap assessment based on the weak points highlighted by a data maturity assessment.

DAMA focus areas	Data Fabric
Data Governance	Supports decentralized, federated, and centralized Data Governance
Data Quality	A proactive automated approach to Data Quality that includes Data Observability. It addresses data at rest and in flight.
Data Architecture	Distributed, metadata and event-driven data architecture with a composable set of tools, technologies, and data storage solutions

DAMA focus areas	Data Fabric
Data Modeling and Design	Semantic data modeling that leverages a Metadata Knowledge Graph to create a connected data ecosystem
Data Security	Automated and embedded Data Privacy, Data Protection and Data Security
Data Integration and Interoperability	Automated Data Integration and interoperability driven by Data Governance and a semantic knowledge layer
Metadata	Active metadata approach to the collection and integration of technical, business, operational, and social metadata

Figure 5.2 – DAMA's data maturity areas mapped to Data Fabric building blocks

Another critical aspect of implementation is available funding. Nothing is possible unless there is a clear budget and prioritization associated with its real distribution. A framework focused on this is FinOps (`https://www.finops.org/introduction/what-is-finops/`). Funding and cost factors will influence the implementation approach and solution selected. A roadmap that aligns the timeline of activities, implementation approach, and solutions over time is necessary. The solution(s) can evolve to cover a wider scope of gaps and data maturity focus areas. Specifically, in the case of a Data Fabric architecture, a phased approach is possible due to its composable architecture.

Let's summarize what we've learned in this chapter.

Summary

A data strategy captures an organization's starting point, journey, goals, solutions, and target destination in order to become a data-driven enterprise. Organizations first need to recognize and understand how and why to manage data as an asset and a Data Product.

In this chapter, we discussed the value of a data strategy document enabling organizations to have a vision and plan to achieve profitable revenue and cost savings by leveraging data. We provided guidance on creating a data strategy and emphasized that it must have a business point of view across people, processes, and technology. We highlighted three popular data maturity frameworks (DMBOK, DCAM, and Gartner's) and recommended completing a data maturity assessment as input for a data strategy. We also covered the implementation of a data strategy document and offered an example that aligned DAMA's data maturity areas with a Data Fabric. Data management, Data Governance, Data Privacy, and Data Security are key focus areas that should be covered by the data strategy. As the business evolves, the data strategy needs to be updated to stay relevant and to ensure continuous progress toward business value.

In the next chapter, *Chapter 6, Designing a Data Fabric Architecture*, we will introduce a logical Data Fabric architecture.

Part 3:
Designing and Realizing Data Fabric Architecture

To create and execute Data Fabric architecture, an understanding of what the design entails, with an established reference architecture and technical direction, is necessary.

This third part has five chapters that introduce the logical architecture of Data Fabric and outlines a technical data architecture to enable implementation. It starts by providing an understanding of what enterprise architecture is and the role Data Fabric plays in it. A deeper dive is then carried out into each of the three layers that compose Data Fabric: Data Governance, Data Integration, and Self-Service. The kinds of tools to enable implementation are listed, and this part ends with a list of industry best practices to carry out.

By the end of *Part 3*, you will know how to design the three layers of Data Fabric architecture. You will also understand the guardrails and principles that need to be embedded into the design. You will know the kinds of tools you can use to drive implementation. Finally, you will have a list of industry best practices and understand why you should follow them.

This part comprises the following chapters:

- *Chapter 6, Designing a Data Fabric Architecture*
- *Chapter 7, Designing Data Governance*
- *Chapter 8, Designing Data Integration and Self-Service*
- *Chapter 9, Realizing a Data Fabric Technical Architecture*
- *Chapter 10, Industry Best Practices*

6

Designing a Data Fabric Architecture

A Data Fabric design represents the evolution of lessons learned from other data architecture approaches that have taken a crack at resolving critical pain points in managing digital data. A Data Fabric architecture design is based on recognized data principles, best practices, and architecture elements. Three foundational layers compose Data Fabric architecture – Data Integration, Data Governance, and Self-Service. Data Integration handles the development and processing of data; Data Governance establishes controls that govern, protect, and secure data; and Self-Service facilitates data sharing. All three layers work together to provide the right balance of control to attain, govern, and share data effectively.

In this chapter, I'll introduce a logical Data Fabric architecture and a point of view on the necessary architecture components. You will navigate through the various layers that compose Data Fabric architecture using simple concepts and rationale. I'll start by introducing the role of data architecture within enterprise architecture and its relevance to Data Fabric. I'll offer a bird's-eye view of the responsibilities of each Data Fabric architecture layer and discuss established architecture principles applied. In the proceeding chapters, I'll dive deeper into the design of a Data Fabric architecture.

In this chapter, we'll cover the following topics:

- Introduction to enterprise architecture
- Data Fabric principles
- Data Fabric architecture layers

Introduction to enterprise architecture

Let's start by defining what enterprise architecture is. **Enterprise architecture** represents an architectural approach to how **Information Technology (IT)** supports an organization's business strategy and capabilities. The objectives are to align an organization's strategic business goals and drivers with a well-thought-out architecture that follows industry principles. **The Open Group Architecture Framework (TOGAF)** – https://www.opengroup.org/togaf – is a highly recognized and prestigious architecture methodology that provides an approach to creating enterprise architecture. It was established in 1995 and has been continuously updated to stay abreast with the latest industry trends, best practices, and principles. TOGAF defines enterprise architecture as follows:

> *...It can be used to denote both the architecture of an entire enterprise, encompassing all of its information systems, and the architecture of a specific domain within the enterprise. In both cases, the architecture crosses multiple systems, and multiple functional groups with the enterprise.*

Figure 6.1 depicts an enterprise architecture made up of four kinds of architecture layers (business, data, application/system, and technology):

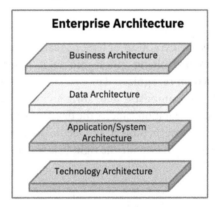

Figure 6.1 – Enterprise architecture

Each has its own level of focus. The following is a summary of each according to TOGAF (http://www.opengroup.org/public/arch/p1/enterprise.htm):

- **Business architecture**: Focuses on providing an understanding of an organization's business model, strategy, business processes, and products or services. It centers around visually depicting the business goals of an organization, business scenarios, and core dependencies. Business architecture provides context into the use cases and value chains a data architecture such as Data Fabric architecture must support.

- **Data architecture**: Entails the data layer that constitutes data and information structures across the organization at both a logical and physical level. This includes all facets of data management, including data storage, data access and movement, Data Governance, metadata, and data operations. Data Fabric falls under this type.

- **Application or system architecture**: This architecture is specific to an application or system and its inner workings. How is the architecture structured? What are its integration points and interactions with other applications and systems, including data architecture? What business processes does it support and how does it tie back to the business architecture?

- **Technical architecture (middleware or infrastructure architecture)**: The key objective of this architecture is to facilitate the deployment and implementation of the application and data architecture. Technology choices (software and hardware) are determined in this architecture to realize the capabilities across both data and application architecture domains.

In the next section, we will discuss data architecture and system architecture and offer examples of each.

Types of enterprise architecture

In the previous section, we discussed the different levels that compose enterprise architecture (business, data, application/system, and technical). The sequence of when to design each architecture is typically the following: business → data → application/system → technical. However, the reverse is true for implementation. Once all the designs across the levels of enterprise architectures are in place, technologies are selected to start implementation. Implementation typically happens from the bottom up, starting with a technical architecture. However, this isn't a strict rule, as implementation largely depends on the requirements. Are you starting from a blank slate, or are you modifying an existing architecture? In some cases, changes to an application/system architecture don't always translate to changes in data architecture.

From a design standpoint, you start by creating a business architecture that addresses the organization's strategy and goals with necessary business capabilities. After business architecture comes data architecture, which focuses on all data management activities across the organization. Third on the list is the application/system architecture, which encapsulates the applications and systems that operate and interact with the data architecture. Finally, the last one is the technical architecture. It is concerned with the selection of technologies and tools to enable implementation. *Figure 6.2* summarizes the design sequence of the levels of enterprise architectures.

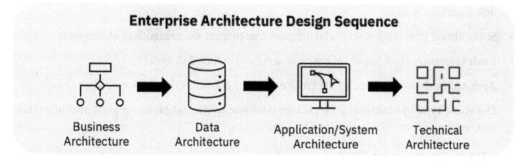

Figure 6.2 – Enterprise architecture sequence

In the following sections, I have listed examples of business, data, and system architecture to provide more context.

Business architecture

Business architecture is a view from a business standpoint of the capabilities necessary to support an organization's goals and objectives. An example of business architecture can include many focus areas, such as the following:

- Customer relationship management
- Product and services management
- Employees management
- Financial

Data architecture

Application/system architecture design can influence data architecture design, such as in the case of **Domain-Driven Design (DDD)** – `https://martinfowler.com/bliki/DomainDrivenDesign.html` – which focuses on developing software based on the context of a bounded domain; in this case, an application/system architecture inspired by applying domain driven design principles to data management such as in Data Mesh. Data Mesh, as discussed in *Chapter 3*, is a federated organizational approach to data management via business domains. This has led to several modern data transformation initiatives that are breaking apart monolithic data architectures in favor of modern decentralized architectures. Monolithic systems are candidates for disintegration in order to remove tight coupling in data and its relationships within the system. In the book *Software Architecture: The Hard Parts*, the authors have listed the following data disintegrator drivers:

- **Change control**: High degree of impact on services from required database updates. This can have a huge impact if a simple change to a table schema causes numerous broken business processes across several dependent services.
- **Connection management**: Database overload from multiple connections needed by distributed services.
- **Scalability**: Limited database scalability, not able to meet the demands of all services.
- **Fault tolerance**: High impact of multiple services from data downtime.
- **Architectural quanta**: Data architecture negatively constricts the software architecture.
- **Database type optimization**: Necessary database optimization to support multiple data system types.

The trend in data management today is to move toward decentralized data management inspired by the popularity of microservices system architecture. However, with these data management approaches, there are considerations that need to be taken. Much of the rigor and control offered by monolithic data architectures, such as in a single relational database, includes the ability to enforce **Atomicity,**

Consistency, Isolation, and Durability (ACID) based transactions. ACID avoids transactions leading to data inconsistency and integrity issues.

In a decentralized model, ACID is no longer enforced by a single database used by different services, which on the positive side creates data consistency. An individual service interacting with a database can still establish that consistency via ACID. However, the likelihood of having one service that handles all possible data write requests in a distributed architecture is unlikely. This architecture typically relies on eventual consistency, which is the notion that eventually, data will be in sync. This comes with its trade-offs. On the positive side, it offers flexibility with increased performance, scalability, and speed. However, data integrity, quality, and the necessary governance need to have a dedicated focus from end to end by the data architecture.

There are justifiable reasons to integrate data and create a monolithic database, such as to support a small business or contained business application. An alternative approach to the physical integration of data within a system is to leverage a Data Fabric architecture that supports decentralized, federated, and traditional centralized/monolithic architectures leveraging metadata as the primary driver to achieve Data Integration. Data Fabric is a type of data architecture that adds an abstraction layer above data storage to enable Data Integration. The following are examples of other types of data architectures:

- Data warehouse
- Data lake
- Data lakehouse
- Event driven
- Data Mesh

A data architecture such as Data Fabric is flexible and works with other data architectures such as data warehouse, data lake, data lakehouse, event driven, and Data Mesh.

Application/system architecture

Application/system architectures are concerned with the degree of service coupling or how tightly coupled services are with one another. The following are some examples of system architectures:

- Microservices
- Event-driven – this architecture may apply to both system and data architectures
- Monolithic

Depending on the trade-offs decided in data and application/system architectures, it will create demand and reliance on when and how a data architecture needs to kick in to address these needs. As an example, in the case of service-to-service communication, whether asynchronous or synchronous is used, it might create data integrity challenges with potentially lost data.

In the next section, we will dive into important principles that need to be designed into a Data Fabric architecture.

Data Fabric principles

Let's start with the objectives of data architecture. It pivots on how IT is used to manage the life cycle of data. It addresses data from the point it is created, stored, transformed, cleansed, and consumed to its end of life. Data architecture is the foundation and strategy that provide the backing to achieve data-driven initiatives. So, what are the best practices and principles that make a good working data architecture? It will need to exhibit more positive trade-offs when it comes to architecture decisions than negatives, as called out by the authors of *Software Architecture: The Hard Parts*:

> *...least worst collection of trade-offs – no single architecture characteristics excels as it would alone, but the balance of all the competing architecture characteristics promote project success.*

A best practice is to document architecture decisions both business and technical in the form of Architecture Decision Records (ADRs) (Source: Software Architecture: The Hard Parts by Neal Ford, Mark Richards, Pramod Sadalage, and Zhamak Dehghani). An architecture decision includes details such as options considered, and impact of decision. This helps reinforce the direction of a data architecture and solidifies the necessary support to move forward.

> **Note**
> As technology matures, business models evolve. As the market continues to change, so will the definition of what represents a good data architecture.

At this present time, how I define quality and good data architecture is it should achieve the following outcomes:

- Support top data management approaches – currently, those are decentralized and centralized.

- Ensure flexibility in managing and enforcing Data Governance based on different models – currently, those are federated, decentralized, and centralized traditional Data Governance.

- Have high-quality, fit-for-purpose data with embedded governance and security. It should enable frictionless, high-scale data sharing with a value proposition that supports the life cycle of data from creation to end of life.

- Effectively and efficiently manage data and information by adopting best practices, architecture principles, and successful frameworks that treat data as an asset and manage data as a product.

- Offer a level of guarantee in its feasibility to address the demanding needs for big data across the four Vs – volume, variety, velocity, and veracity.

- Have a robust architecture that can evolve with the future needs of the industry without major architecture rework.

In the next section, we will review the data architecture principles of Data Fabric design.

Data Fabric architecture principles

In previous chapters, we introduced DataOps, and Data Mesh principles focused on achieving business value from data. We will introduce additional principles from TOGAF (https://pubs.opengroup.org/togaf-standard/adm-techniques/chap02.html#tag_02_06_02), and Google (https://cloud.google.com/blog/products/application-development/5-principles-for-cloud-native-architecture-what-it-is-and-how-to-master-it) that represent the foundation of a good data architecture.

Following is a collection of foundational data architecture principles required in a Data Fabric design:

- Data are assets that can evolve into Data Products (TOGAF, Data Mesh, and DataOps Manifesto)
- Data is shared (TOGAF)
- Data is accessible (TOGAF)
- Data product ownership (TOGAF and Data Mesh)
- Common vocabulary and data definitions (TOGAF)
- Data security (TOGAF)
- Interoperable (TOGAF)
- Always be architecting (Google)
- Design for automation (Google)

These principles capture the essence of what is necessary in a modern Data Fabric architecture. DataOps has a strong correlation in a Data Fabric architecture and will be discussed in the next section.

Let's dive into each principle to understand its relevance in Data Fabric design.

The principle of data are assets that can evolve into data products

The biggest realization in today's digital data era is that treating data as a byproduct resulting from a business process or system leads to a slew of issues that ultimately creates a loss in profit and financial return. This principle represents a culture change where treating data as an asset represents governed, trusted, accurate, and secure data. Managing data as a product represents a transition where assets have active product management across their life cycle from creation to end of life, have an intended value proposition and are enabled for high scale data sharing. Data must have the following usability characteristics: discoverability, addressability, understandability, trustworthiness, native accessibility, interoperability, and independently valuable. Data must be proactively monitored with automation, governance, and quality control best practices to achieve critical decision-making.

The *data are assets that can evolve into data products* principle is inspired by the TOGAF, Data Mesh, and DataOps Manifesto (`https://dataopsmanifesto.org/en/`) principles, listed here:

- Data is an asset (TOGAF)
- Manage data as a product (Data Mesh)
- Quality is paramount (DataOps Manifesto)
- Value working analytics (DataOps Manifesto)
- Monitor quality and performance (DataOps Manifesto)
- Analytics is manufacturing (DataOps Manifesto)

The principle of data is shared

Data sharing unlocks siloed distributed data. It enables decision-makers to leverage untapped data. Data sharing across different organizations and business data domains is a necessity. It's important to recognize the necessary rigor on data to manage shared access and implement the right level of controls to prevent data duplication and unnecessary data transformations for the purpose of addressing targeted use cases. It is equally as important to address the diversity of data sharing across different data systems, hybrid cloud environments, and technologies. This is a key principle in not only Data Fabric but also any other mature data management approach, supporting and empowering data producers to share data and making it easy for a diverse set of business and technical data consumers to access data. This should take into account the principle "Data are assets that can evolve into Data Products" as not all assets are created with an intention to share at high scale.

Best practices in data engineering and data processing are applicable here, such as in the development and deployment of analytical data pipelines. Managing data as code is a best practice that enables flexibility and a high degree of automation that achieves efficiency. However, a non-programmatic approach using a low-code or no-code tool is another option. Flexibility in supporting both a programmatic and non-programmatic approach in the development and deployment of assets/data products should be in place. What's key is irrespective of which approach is taken (programmatic or non-programmatic), there needs to be a common layer that ties all the data together and makes sense of it to enable data sharing. Therefore, this principle is closely aligned with the interoperability principle.

The principle of data is accessible

As defined by TOGAF, this involves the ease with which users obtain information. Data accessibility must support different modes of access tailored to different user roles whether in business or technical roles. This should keep in mind the typical modes of access necessary by role; for example, data engineers or data scientists might prefer more intricate formats, such as Parquet files and the use of APIs, as opposed to business analysts, which may prefer to have data delivered in the data visualization file format of choice. Data access must have embedded governance that addresses Data Privacy and Security.

The principle of data product ownership

According to the principle *data are assets that can evolve into data products*, data must be actively managed. A data product owner is accountable for the overall quality, security, privacy, and success of data. They manage consumer requirements, the data life cycle, and the operationalization of data and ensure it delivers business value to data consumers. It is a different role from a data steward, which has broader responsibilities, such as defining and implementing data policies/standards across an organization. A data product owner focuses on quality control, manages the entire life cycle of an asset(s) evolving into a Data Product and are accountable for high scale data sharing. Depending on an organization's makeup the role of a Data Product owner maybe fulfilled by other business roles in the organization. In the book *Data Mesh*, Dehghani describes data product owners as follows:

> *...facilitate making decisions around the vision and the roadmap for the data products, concern themselves with the continuous measurement and communication of their data consumers' satisfaction, and help the data product team improve the quality and richness of their data.*

The principle of common vocabulary and data definitions

Business semantics that adequately describe data in an organization and its relationships are key. The goal here is to eliminate ambiguity in multiple teams referring to the same data using different semantics. There are several best practices in how to achieve this depending on the maturity of the organization and their data modernization journey. A traditional best practice for a central Data Governance organization is to establish a canonical enterprise model that captures the business vocabulary for the organization to establish standardization. The latest best practice is to support a federated Data Governance organization where each business domain has a common vocabulary established, although it has strong interoperability standards that enable its sharing across diverse business domains.

The principle of data security

Data needs to be secured and protected according to Data Privacy laws and regulations. Data needs to be classified according to the level of enforcement it requires, such as sensitive, private, or other classifications. The necessary policies need to be in place to determine the course of action necessary to drive security and privacy. Data access is determined by data access policies applicable to a data consumer based on the classification of data. It needs to be designed or integrated with data access methods. As TOGAF highlights, the necessary data security safeguards need to be in place to restrict access as needed based on the sensitivity labeling.

The principle of interoperability

To enable data sharing, semantic metadata that provides the necessary understanding to share and consume data is needed in addition to ensuring the right technical specifications, making it simple to get data from a source and use it in a target setting. A semantic metamodel that captures data and links its relationships offers an understanding of how to use and leverage data. There needs to be a set of standards on the expected level of metadata collected to register and define data.

From a technical standpoint, the use of APIs to enable data use is an example of how interoperability can be supported. However, a specific format and standards that stipulate the schema of the API are still needed. This creates commonality that will enable finding, understanding, and sharing data more easily. This is especially important for distributed data architectures. An example is in Data Mesh, where it aims to resolve data silos via a distributed organizational approach. However, if Data Mesh is not implemented correctly following an interoperability principle, it can itself contribute to data silos.

The principle of always be architecting

Google Cloud's principle of always be architecting highlights the importance of a data architect's role. You're never truly done. Technology is changing rapidly, and new approaches and best practices are emerging every day as we gain more experience and learn what works and what doesn't. New business requirements can drive changes necessary in the existing data architecture. It's important to always learn and stay up to date with the latest best practices, technologies, and industry trends to assess more effective ways of improving existing data architectures. Leverage lessons learned to refine and improve a data architecture. The DataOps Manifesto principle of embrace change aligns with this principle.

The principle of design for automation

Google's design for automation principle instills the need for high automation in data management. The goals are to expedite cycle times in data collection, development, and deployment to deliver data quickly and achieve data monetization. Continuous Integration and Continuous Delivery, data monitoring, and Infrastructure as code practices are advocated. The following are DataOps Manifesto principles that align:

- Analytics is code

- Orchestrate

- Improve cycle times (DataOps Manifesto)

Now that we've covered the principles of Data Fabric design, let's move on to the different layers of Data Fabric architecture.

Data Fabric architecture layers

Data Fabric architecture follows the nine principles discussed in the previous section. These principles establish the bedrock for a Data Fabric architecture that addresses data silos, enables data democratization, and creates a connected and intelligent data ecosystem of trusted, secure, and reliable data that supports data producers and data consumers. In *Chapter 1, Introducing Data Fabric*, we discussed Data Fabric design as having three building blocks:

- Data Governance

- Data Integration

- Self-Service

Let's represent these building blocks as layers in a Data Fabric architecture with specific responsibilities and supporting components:

- **Data Governance** enables active Metadata Management and automated life cycle governance. Its knowledge layer is represented by a Metadata Knowledge Graph.

- **Data Integration** handles inbound and outbound data management, data processing, and data engineering. This layer relies on active metadata and governance capabilities in the Data Governance layer.

- **Self-Service** executes data democratization by supporting data producers and data consumers to enable governed data sharing. This layer also relies on active metadata and governance capabilities in the Data Governance layer.

Data Fabric architecture supports distributed data across diverse storage systems in a hybrid and multi-cloud landscape. It handles operational and analytical data. **Operational data** is typically used in day-to-day operations critical to the success and reliability of the business. **Analytical data** gathers business and data intelligence to derive insights and make business decisions. *Figure 6.3* depicts data producers and data consumers interacting with a Data Fabric architecture consisting of three layers.

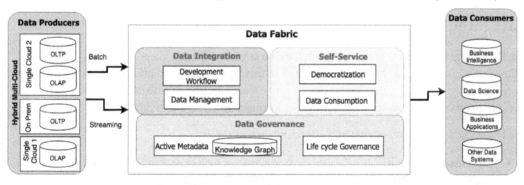

Figure 6.3 – Logical Data Fabric architecture

In the next section, we will discuss in more detail the three layers of Data Fabric architecture.

Data Governance

The Data Governance layer determines the right course of action to govern and enforce data policies. It's focused on ensuring Data Governance is in place to execute the principle of *data are assets that can evolve into data products* and establish Data Security and a business vocabulary to enable its use. It gathers and analyzes all metadata extensively into its knowledge graph. An analogy would be the human body consulting with the brain to know what the next action should be. The Data Governance layer works with the Data Integration and Self-Service layers to ensure the right level of Data Governance is applied to incoming and outgoing data in a Data Fabric architecture.

The following are two logical components of the Data Governance layer:

- **Active metadata** pushes the envelope from traditional Metadata Management or passive metadata to modern Metadata Management. It's a key differentiator when compared to other styles of data architectures. It represents an intelligent set of capabilities that manage the collection, processing, and analysis of metadata across the entire Data Fabric platform. A Data Fabric's knowledge graph represents the storage of a 360-degree metadata view of data across the entire Data Fabric architecture. Often, its underlying data layer is a graph database that pulls all the pieces of its understanding together across disparate data systems, including other metadata repositories and tools. The role of the active metadata component in the Data Governance layer is to study relationships, make correlations, derive new intelligence, and execute recommendations based on that knowledge.

- **Life cycle governance** is paramount to a Data Fabric architecture. It embeds and automates Data Governance capabilities focused on Data Privacy, Data Protection, Data Security, Data Quality, Data Lineage, Master Data Management, and Metadata Management. It applies these capabilities in an automated fashion at every stage in the life cycle of data as it navigates the Data Integration and Self-Service layers of a Data Fabric architecture.

Data Integration

The Data Integration layer is concerned with the data engineering and data processing operations that follow DataOps principles. Metadata generated in this layer is captured and processed in the Data Governance layer to build a 360-degree view of active metadata. The objectives of the Data Integration layer are to enable the development, testing, deployment, and delivery of data.

The following are two components of the Data Integration layer:

- **Data management** handles data processing activities such as data ingestion, data storage, data transformation, data preparation, and data provision. Data processing can take place from **Online Transactional Processing** (**OLTP**) systems and **Online Analytical Processing** (**OLAP**) systems across a hybrid multi-cloud landscape. Data ingestion and integration activities should support diverse methods (batch and streaming), querying, and technical processing approaches.

- The **development workflow** applies DataOps to development and deployment activities. Continuous integration and continuous deployment practices with quality control are necessary to enable Data Integration with high automation across disparate data sources.

Self-Service

The third layer is *Self-Service*, which is concerned with supporting data product owners and data consumers. Data product owners create data that is accessible and shareable with product management capabilities that follow the principle *data are assets that can evolve into data products*. Data consumers represent a diverse set of technical and business users that are empowered with Self-Service access and consumption of data. The Self-Service layer relies on the Data Integration layer to develop and deliver high-quality data, and the Data Governance layer to govern data during access and delivery.

The following are two components of the Self-Service layer:

- **Data democratization** focused on supporting the data product owner in the onboarding and operationalization of data. This involves the publication and sharing of data to enable its discovery and use by a data consumer.

- **Data consumption** supports a data consumer's needs in their discovery, understanding, access, delivery, and composability of data. Data consumption supports a diverse set of data consumers by enabling the delivery of fit for purpose, value driven, and consumable data across different delivery streams.

Let's summarize what we've covered so far in this chapter.

Summary

A Data Fabric design represents a data architecture that is part of enterprise architecture. It's critical for a data architecture to align and support the business capabilities designed as part of the business architecture for an organization. To design Data Fabric, it must follow nine recognized industry principles, which are data are assets that can evolve into data products, data is shared, data is accessible, data product ownership, common vocabulary and data definitions, data security, interoperability, always be architecting, and design for automation. These principles provide the necessary guardrails in its design to ensure it provides business value while following successful practices.

In this chapter, we provided an introduction to a logical Data Fabric architecture consisting of three layers: Data Integration focused on inbound and outbound data management; Data Governance, ensuring the right level of data integrity, security, and trust are in place with a rich active metadata layer; and the Self-Service layer, which is concerned with data democratization. We discussed how these layers form the foundation of a Data Fabric architecture design.

In the next two chapters, *Chapter 7, Designing Data Governance*, and *Chapter 8, Designing Data Integration and Self-Service*, we will dive deeper into the layers of Data Fabric architecture.

7
Designing Data Governance

The job of the Data Governance layer in a Data Fabric architecture is to govern, protect, secure, and enable the discovery and understanding of data. A Data Governance layer in a Data Fabric builds a connected data ecosystem with metadata, embedding Data Quality, Data Lineage, Data Security, Protection and Privacy, and **Master Data Management** (**MDM**) into every phase of the life cycle of data. It is driven by an active metadata-infused framework that enables enforcement and automation.

In this chapter, we'll review the design of the Data Governance layer in a Data Fabric architecture. We'll discuss its metadata-driven and **Event-Driven Architecture** (**EDA**) patterns. Finally, we'll step through how the Data Governance layer is applied to the data life cycle.

In this chapter, we'll cover the following topics:

- Data Governance architecture
- Metadata as a service
- The Data Governance layer
- Data Fabric's governance applied

Data Governance architecture

The Data Governance layer establishes the foundational data architecture for a Data Fabric, which both the Data Integration and Self-Service layers rely on. The following are two key architecture patterns in the Data Governance layer:

- Metadata-driven
- Event-driven

Let's discuss each of them in detail.

Metadata-driven architecture

Metadata in data management is defined classically as *data about data*. It provides context and information relevant to data so that it can be understood and used effectively. Metadata collection has a long history, dating back to the late 1900s (`https://www.dataversity.net/a-brief-history-of-metadata/`). The reliance on and use of metadata since then have increased drastically throughout the years with digital data management. Companies such as Google and Adobe were pioneers in tapping into the value of leveraging metadata to create a world of discoverable and usable data at a grand scale to realize business value. In the days of Ralph Kimball, an author and thought leader on data warehouse architectures, he leveraged metadata to enable data discovery and data management. This gave rise to data catalogs in the 1990s. **Data catalogs** are metadata repositories that capture and organize metadata. They create an inventory of assets, associated data storage, and other details to enable relevant and organized metadata within data usage across an organization.

Today, data catalogs are extremely relevant to the discovery and sharing of data. They have matured from an inventory tracker of data assets or a simple data dictionary into managing Data Governance across large-scale organizations where metadata is treated as a service. Metadata is leveraged to support effective data management in diverse data storage solutions and automate Data Governance. Shirshanka Das calls this a second-generation architecture (`https://engineering.linkedin.com/blog/2020/datahub-popular-metadata-architectures-explained`). One of the most important and critical principles a data catalog solution targets is the dominant use of metadata to drive data management at a large scale.

Four types of metadata were introduced in *Chapter 1*:

- **Technical metadata** – The data structure based on where the data resides. This includes details such as the source, schema, file, table, and attributes.

- **Business metadata** – This provides business context to the data. This includes the capturing of business terms, synonyms, descriptions, data domains, and rules surrounding the business language and data use in an organization.

- **Operational metadata** – Metadata produced in a transactional setting, such as during runtime execution. This includes data processing-based activities such as pipeline execution with details such as the start and end time, data scope, owner, job status, logs, and error messages.

- **Social metadata** – Social-based activities surrounding the data. This metadata type has become more prominent as it focuses on putting yourself in a customer's shoes and understanding what they care about. What are their interests and social patterns?

Metadata is produced by a range of databases, metadata repositories, operational pipelines, data science, and data visualization tools. Metadata, just like data, is extremely fragmented and spread across many repositories. Metadata collection can take place at different times and levels of granularity and for different purposes without any standardization or best practices. *The big question is, what if we zoomed out and connected technical, business, operational, and social metadata to data as it is created, ingested, integrated, and consumed, up until its end of life?* That is, a world where metadata is collected across distributed data storage, metadata repositories, data processing, and analytical tools in a hybrid cloud landscape. This is how metadata-driven architecture works.

In the next section, we will introduce Event-Driven Architecture. The principles of metadata-driven architecture and EDA work together when utilizing active metadata in a Data Fabric architecture.

EDA

EDA represents loosely coupled microservices that typically communicate asynchronously. It borrows many of the principles of a microservices architecture. Synchronous communication can also be leveraged such as in a service-oriented architecture. A best practice, however, is to achieve scalability by communicating asynchronously in an EDA. EDAs have been around for many decades. *Adam Bellemare*, in his book *Building Event-Driven Microservices*, highlighted the following:

> *"...event driven architectures are nothing new. In fact, they were very popular during the days of building Service Oriented Architectures. What has changed over time is the demand for managing big data, with high scalability and speed."*

Today, the demand is to provide access to the latest and highest-quality data in order to take immediate action. This is what gives you the edge of agility and the competitive advantage to keep up with business demands. An EDA exchanges business events between a producer and a consumer, leveraging an event broker. Only services interested in specific types of business events receive them. This offers a level of decoupling that is highly advantageous when compared to point-to-point communication across services.

An EDA operates using a listener model. Events are populated by a producer based on a change in state for a business event. An example of this is a customer creating an order or a new Data Privacy classification for a customer. Based on this change in state, there are listeners waiting to process the event. Asynchronous communication is the ability to send a message without a service having to wait for a response. The event itself is a notification vehicle but can also hold the actual data content and serves as a repository for events. EDA operates based on eventual consistency, which means that the data will be eventually processed and there are no guarantees of when the processing happens. *Figure 7.1* showcases an EDA interaction model.

Figure 7.1 – EDA

It starts with a producer creating an event, an event broker then receives the event, and a consumer subscribed to the relevant events receives them. An event broker is a middle-ground component. It receives events, stores them, and distributes them to all data consumers.

While the focus here is discussing an EDA, a Data Fabric architecture can also have dependencies in microservices and API-based architecture as well.

> **Microservices versus API architecture**
>
> A services-based architecture is the exact opposite of a monolith-based architecture. It represents loosely coupled decentralized services. It is focused on breaking down or modularizing services into smaller containable pieces of code that can be developed, tested, and deployed independently.
>
> An API architecture is a type of software architecture that enables services to communicate with one another, usually in a synchronous manner via a request-and-response structure. It provides an abstraction layer via an API specification that enables other services to call it and receive information.

Some of the benefits of an EDA are as follows:

- Latest and greatest data in real time
- Takes action instantaneously based on business events
- Decoupled communication between producers and consumers
- Focused data sharing
- Scalability – no point-to-point communication via its loosely coupled approach

In the next section, we will discuss how to leverage both architectures, EDA and metadata-driven, in a Data Fabric.

Metadata as a service

Metadata as a service needs to be designed into a platform to drive automation and enable data sharing, governance, and Data Integration. The collection of metadata needs to take place throughout the life cycle of data, across all data management functions such as the development and deployment of data, data ingestion, Data Integration, and data consumption until its end of life. Metadata collection at every stage of the data life cycle enables us to make intelligent decisions to increase business value. In the following section, we will dive into some important points about metadata collection.

Metadata collection

To enable data consumption and enforce the right level of Data Governance, data must be discoverable and understood. This starts by collecting metadata wrapped around data. Metadata should be leveraged to create a 360-degree view that represents everything there is to know about assets, whether data assets, data related assets, and Data Products. There is an enormous amount of effort made in the market today to define Data Products. One source of inspiration at the time of writing this book is the EDM Council (`https://edmcouncil.org`) who are actively defining Data Product, and many other trending industry topics.

What are Data Assets and Data Products?

The definition I will offer is one that is directly in line with the Data Fabric principle introduced in this book - "Data are Assets that can evolve into Data Products." Data Assets represent Enterprise data that is curated and cataloged. Data Assets should be governed although aren't always created with the intention of sharing at high scale, therefore may not have ongoing quality control. Data Products are intentionally created to enable high-scale data sharing, have a specific business value proposition, and can contain several data assets and data-related assets.

Assets become Data Products when they are: 1) operationalized 2) actively data product managed 3) Self-Service. Operationalize entails the packaging of assets into business-ready Data Products, Data Contracts are established, and data delivery channels are set up. Active data product management applies a focus on deriving business value from data across its entire life cycle. Self-Service enables data to be easily accessible directly by data consumers.

The following list is a starting point of metadata repositories sourced from Piethein Strengholt in his book Data Management at Scale. Typically, these metadata types live in separate repositories. Strengholt advocates in his book that metadata is the driver that enables data management at scale, which I completely agree with. It's the reason why it's a key driver in a Data Fabric architecture. I have expanded his original list to add a few more. This provides a baseline of repositories to build a framework for metadata collection as part of the Data Governance layer in a Data Fabric. The objectives are to catalog and publish metadata surrounding all aspects of Data Assets, Data related Assets, and Data Products:

- Application, database, and asset repositories
- A **List of Golden Sources** (**LoGS**) repository
- Data product, data sharing, and entitlement/data contract repositories
- Repositories for the conceptual, logical, and physical data models, including metamodels
- Repositories for the messages and interface models
- Repositories for the crowdsourced context
- A lineage repository
- A security policy repository
- A Data Quality repository
- A data classification repository
- A business glossary repository
- Data orchestration and data processing repositories
- Data science and business intelligence repositories

In the next section, we will discuss how to take metadata that lives in distributed metadata repositories and integrate it to build the knowledge layer of a Data Fabric.

Metadata integration

In the previous sections, we established metadata-driven architecture and EDA as the architecture styles that define the Data Governance layer of a Data Fabric architecture. We also went through metadata collection. The next step is to integrate technical, business, operational, and social metadata and establish relationships that create an intelligent knowledge graph. A Metadata Knowledge Graph connects various nodes with defined relationships at different levels of granularity collected at every stage in the life cycle of data. This is a key enabler of high-maturity decision-making in an organization. In the next section, we will dive into what composes a Data Fabric's Knowledge Graph.

A Data Fabric's Knowledge Graph

Let's explain the Knowledge Graph by first talking about the human brain. How the brain works is fascinating. According to this article – `https://www.hopkinsmedicine.org/health/conditions-and-diseases/anatomy-of-the-brain` – the brain heavily relies on signals it receives throughout the body. These signals enable the brain to control different aspects of the body. Different parts of the brain focus on different things, such as the frontal lobe, which enables decision-making, and the parietal lobe, which understands relationships and identifies objects. This makes total sense, as when you think about how we process people, events, objects, and relationships, that is what enables us to correlate findings and make decisions on this intelligence.

I usually like to take a step back and look at the big picture to identify trends and patterns in specific situations to equip myself to decide the best course of action. I think, to some degree, all of us do this subconsciously and effortlessly in everyday life situations. If you buy food at a restaurant and get food poisoning, you most likely won't ever buy food from that restaurant again. You might even decide to post a negative review as a result of this event. As simple as this example is, it shows the power of relationships and connecting the dots at varying degrees of available information at different points in time. The same applies to how the Metadata Knowledge Graph in a Data Fabric works.

Let's start by defining a Metadata Knowledge Graph.

Metadata Knowledge Graphs

According to *Jesus Barrasa et al.* in their book *Knowledge Graphs*, they "...*are interlinked sets of facts that describe real-world entities, events, or things and their interrelations in a human-and machine-understandable format*". **Metadata Knowledge Graphs** represent graphs that collect metadata and associated data relationships. They focus on evaluating collected intelligence at a grander scale with the intent to take action through decision-making. *Figure 7.2* illustrates a Metadata Knowledge Graph capable of answering key business questions:

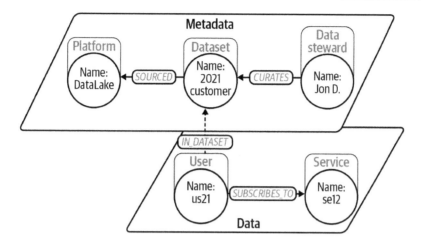

Figure 7.2 – Knowledge Graph depicting metadata and data relationships
(Source: Knowledge Graphs book by Jesus Barrasa et al.)

The power of a Metadata Knowledge Graph is experienced when a business vocabulary is applied to the collection of technical, operational, and social metadata. A set business language allows us to interpret and understand existing relationships in context. There are different topologies available to represent the business language, with varying relationships. Examples are hierarchal taxonomies, ontologies that capture vertical and horizontal relationships, and a traditional enterprise model. Each one has its strengths and weakness and usage largely depend on how a business runs. An enterprise model can serve well for an organization that has done a tremendous amount of work to standardize and define how the entire business communicates across all domains. An ontology provides the most flexibility. It's capable of capturing and storing relationships to reflect how things interrelate in the real world.

In a Data Fabric architecture, a Metadata Knowledge Graph plays a crucial role. It enables the execution of active metadata in the Data Governance layer. It is a repository that collects and stores metadata from various repositories, provides references to physical data repositories, establishes relationships across the metadata, and creates an integrated metadata view. This view enables decision-making that can be acted on with respect to data management; whether this is to initiate the resolution of data defects, avoid the impact of stale data, or implement data policy enforcements to avoid regulatory fines.

Let's dive deeper into a Data Fabric's Knowledge Graph.

Data Fabric Knowledge Graphs

In a Data Fabric architecture, the objective is to leverage metadata as the vehicle that enables effective data management. This creates flexibility in the data layer, where data can physically be located across diverse endpoints representing a hybrid multi-cloud set of environments. The point of using a Metadata Knowledge Graph is to capture and extract metadata at every stage of the data life cycle to know everything there is about it and reference the physical data where it resides. A Data Fabric architecture represents an abstraction layer above the data layer. Some might argue a data catalog can achieve the same thing as a knowledge graph. While a data catalog, in fact, represents a key tool in the implementation of a Data Fabric, as covered in *Chapter 9, Realizing a Data Fabric Technical Architecture*, it is one of many in its composable architecture. A *Metadata Knowledge Graph* is key in a Data Fabric architecture, as it offers a connected ecosystem with a 360-degree view of the metadata associated with your data.

Some of the relationships captured by Data Fabric's knowledge graph are as follows:

- Indexes for where the data physically resides – technical metadata

- Schema and format details that describe the shape and location of data – technical metadata

- Business semantics that describe the context of the data – business metadata

- Data Lineage, which includes data provenance – operational metadata

- Data processing details such as runtime details – operational metadata

- Social relationships such as the social tags, ratings, and data usage popularity – social metadata

The possibilities are vast in terms of the level of knowledge that can be attained.

> **What's the difference between leveraging a knowledge graph versus a Data Governance catalog?**
>
> A data catalog manages an inventory of organizational data assets and in some cases other analytical assets with a heavy focus on Metadata Management. Usually, it has one or more metadata repositories behind the scenes and provides a visual interface/APIs. The data storage system behind a data catalog may or may not use a graph-based data storage system that enables the population of a knowledge graph.

Figure 7.3 shows a Metadata Knowledge Graph and demonstrates how it can work. It captures relationships between a diverse set of metadata repositories and creates references to the physical data residing in distributed data systems.

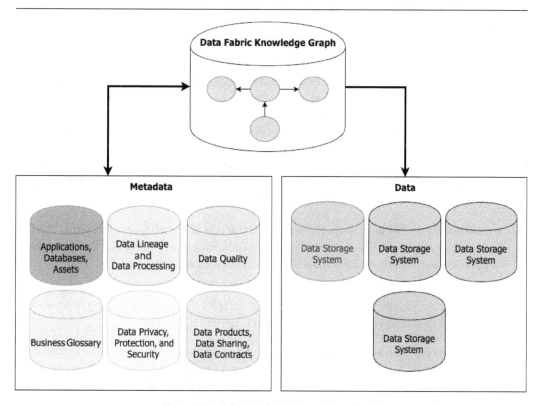

Figure 7.3 – A Data Fabric's Knowledge Graph

Some of the advantages of using a graph-based data system are its agility and flexibility. It can be designed iteratively as needed and updated without major redesign as compared to traditional data storage such as relational databases, which require the data model to be designed upfront. Changes to a relational data model can have a drastic impact on downstream systems and this is a painstaking exercise when it comes to capturing complex relationships. A graph database is often a more suitable technology in that it enables the continuous collection of new and updated metadata. As new data is created, ingested, or updated within a Data Fabric architecture, metadata must be collected along the way. The intent of a Data Fabric's knowledge graph is to build a metadata-connected view of data and its relationships. With this in mind, the majority of the knowledge graph will store metadata, but the relationship of all metadata must always be tied to physical data assets in the form of a reference, as showcased in *Figure 7.3* and in some other cases, the actual data content itself maybe captured.

The power behind the collection of metadata and building a 360-degree view associated with the data is as strong as how fresh the metadata is. If you build a metadata view and make critical business decisions based on it only to find out later that the metadata and data correlations were incorrect, it can lead to significant consequences. While I have extensively discussed the role of a metadata-driven architecture and a knowledge graph, there is one last component that elevates the Data Fabric architecture to the next level of maturity – an EDA. In the next section, we'll discuss the role of EDA in a Data Fabric.

Metadata-based events

In the *A Data Fabric's Knowledge Graph* section, we made an analogy comparing a Data Fabric knowledge graph to the human brain. We discussed the signals the human body sends to the brain to enable the brain to execute control. A **business event** can be compared to the signals sent by the throughout the body with the expectation of the knowledge graph executing action. As new data is created and associated metadata is collected, signals in the form of business events are sent to the event stream. Data consumers process the metadata and act. Examples of business events can range from a new order placed by a customer to the ingestion of data for a particular data domain, or metadata insights collected for a Data Product within a Data Fabric architecture. An event captures all the details associated with the event serving as the source of truth. There are two styles of EDA:

- **A mediated EDA**: This has a mediator component that acts as an orchestrator of events. This is necessary for more complex events that require coordination and involves executing multiple steps that require calling multiple individual business processes.

- **Broker event-driven architecture (without a central mediator)**: This architecture design does not have a central mediator. It's appropriate for less complex use cases where an event is passed to a message broker and any number of consumers retrieve it.

For events to be processed between a producer and consumer services, there needs to be a common communication exchange format to enable interoperability. A common standard format establishes how the services communicate when creating events. The following is a list of best practices to design high-quality events from Adam Bellemare in his book *Building Event-Driven Microservices*:

- In business terms, describe everything that happened in the event – "…*full and total authority on what actually occurred, and consumer should not need to consult any other source of data to know that such an event took place.*"

- Events should represent a single logical event – do not mix different types of events in an event stream, as it may create confusion on what the event is.

- Use the narrowest data types to avoid ambiguity.

- Events need to be single purpose – avoid using type fields that represent general events and not a single business activity, as indicated in *best practice 2*.

- Minimize the size of events – they should be small, well defined, and easily processed. Large events require a second look.

- Avoid events as semaphores or signals – events need to include both data results and event details.

Now, let's discuss how an EDA and a metadata-driven architecture, together with a knowledge graph, establish the Data Governance layer.

The Data Governance layer

The Data Governance layer in a Data Fabric architecture represents different architectural patterns brought together to ultimately create a fabric of connected metadata and data. There are two key components in the design of the Data Governance layer – **active metadata** and **life cycle governance**. Each of these components at a physical level can be made up of several microservices that communicate with one another by leveraging different styles of communication such as APIs and event-based communication. The key architectural pattern in the Data Governance layer relies on metadata. From an implementation perspective, several technologies and tools are involved in this space, such as a data catalog, which leverages a streaming platform to communicate metadata updates. Leveraging an event-based tool for metadata updates ensures the latest most up-to-date metadata is collected.

Figure 7.4 represents the Data Governance layer, consisting of active metadata and life cycle governance. Each of these components has a set of supporting capabilities as listed here. The operational model that the Data Governance layer supports should be flexible in supporting the key ones today, which are centralized, federated, and decentralized.

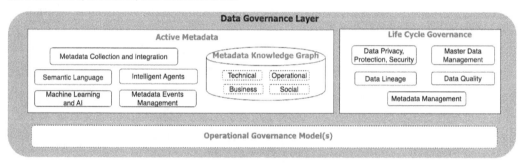

Figure 7.4 – The Data Governance Layer

Let's dive deeper into the three key components: active metadata, life cycle governance, and an operational model.

Active metadata

The key objective with active metadata is to have a connected metadata ecosystem. This has to collect metadata, create knowledge, and then take action based on this knowledge to drive increased performance, trust, and quality in data management. That's the biggest distinction between passive and active metadata.

The active metadata component consists of the following:

- **Metadata Knowledge Graph**: The integrated metadata view captured in a knowledge graph enables the Data Fabric. Metadata comes from various metadata repositories and an ecosystem of tools, databases, and data environments.

- **Metadata collection and integration**: This is the continuous collection, discovery, and integration of metadata from various repositories, which leverage event-based and API-based communication methods. It supports both push and pull methods of metadata collection. Data is processed and stored in the Metadata Knowledge Graph component.

- **Semantic language**: This represents an ontology, taxonomy, or enterprise model that provides context to the data and the relationships in its metadata.

- **Intelligent agents**: Agents represent service(s) that can process, and analyze the metadata, create insights and make decisions based on the intelligence generated in the Metadata Knowledge Graph.

- **ML and AI**: ML and AI technologies that automate and learn as they go to increase the level of maturity in Metadata Management. This can play a role in the auto-classification of data, auto business enrichment, master Data Integration, and Data Quality.

- **Event management**: Managing events represents metadata updates within a Data Fabric architecture.

Figure 7.5 depicts an active metadata example using a metadata-driven architecture/and event-driven architecture.

Figure 7.5 – Example using a metadata-driven architecture/EDA

As data is created, events are generated by the metadata collection service (producer), which are sent to an event stream. The metadata integration service (the consumer) is notified of a new metadata event and processes it by loading it into the Metadata Knowledge Graph. The metadata integration service (the producer) then generates a new event to notify subscribed data consumers of new or updated metadata.

In the next section, we will review life cycle governance.

Life cycle governance

Life cycle governance represents the critical need to embed and automate Data Governance in a Data Fabric architecture. It represents capabilities across important governance pillars that support all phases in the life cycle of data. Life cycle governance works together with a Data Fabric's active metadata component to deliver automated, intelligent, and proactive governance. To achieve successful data management, it must take a three-dimensional approach. Data Governance, similar to the DataOps framework discussed in *Chapter 4*, has people-related, process-related, and technological dimensions.

Data Governance in a Data Fabric architecture must address who, what, where, when, and how Data Governance is kicked into gear across these three dimensions:

- **People**: Make data roles accountable and responsible to manage and enforce Data Governance. This may include roles such as data product owners, data stewards, data security and privacy focal points, data owners, data product managers, and data trustees. This also includes supporting data consumers to make business decisions based on trustworthy and reliable data, such as business analysts, data engineers, data scientists, architects, or executives.

- **Process**: Business and data processes in an organization are driven by the business strategy, given objectives, and the capabilities of an organization.

- **Technology**: Tools and technology must have the right set of Data Governance capabilities, automation, and maturity to properly manage, protect, share, and secure data.

The life cycle governance component consists of capabilities across the pillars introduced in *Chapter 1, Introducing Data Fabric*:

- Data Privacy, Protection, and Security

- Data Quality

- Data Lineage

- Master Data Management

- Metadata Management

Operational models

A Data Fabric architecture should have capabilities that enable its ability to support diverse operational models. This includes the ability to enable different roles across different business domains to take accountability and responsibility in Data Governance. This introduces the need to potentially enforce the federation of metadata across platforms to achieve data sharing and Data Governance. It also requires isolating Data Governance-focused tasks, artifacts, and metadata according to the taxonomy that matches the operational model (central, federated, or distributed).

In the next section, we will review a day in the life of data in the governance layer of a Data Fabric architecture.

The Data Fabric's governance applied

The data life cycle is typically represented by the journey data takes from when it comes into existence to the point at which it is destroyed. As this applies to the Data Governance layer in a Data Fabric architecture, certain specific steps and considerations are taken. Each of the actions taken respects the Data Fabric principles as part of the overall architecture. *Figure 7.6* represents the logical data life cycle phases in which Data Governance must be applied. Data won't always flow across every phase. However, the Data Governance layer must account for every phase with the necessary capabilities to ensure it is properly protected and secured.

Figure 7.6 – Data life cycle phases

> **Note**
>
> The life cycle of data can be represented in different ways with various points of view depending on the scope and objectives. There isn't one standard, as all have their own level of focus. Some might call out data storage as its own phase or use slightly different naming conventions. What's more important as part of this book is to depict the role Data Governance must play in the logical life cycle of data in a Data Fabric architecture.

Let's review a day in the life of data as it is managed by the Data Governance layer in a Data Fabric architecture. How are active metadata and life cycle governance applied to data from **Create → Ingest → Integrate → Consume → Archive and Destroy**?

For the sake of simplicity in following this journey, I have grouped the Data Governance pillars of Metadata Management, Data Lineage, and MDM, except in the Integrate phase.

The Create phase

Data Governance typically starts for existing data residing in a storage system or at *rest*. While this certainly is one of the critical touch points, I would argue the first touch point needs to start during the development of data. This is typically referred to as a *shift-left* best practice (https://www.forbes.com/sites/rscottraynovich/2022/01/13/the-shift-left-is-a-growing-theme-for-cloud-cybersecurity-in-2022/?sh=3e839f5d7ff1). Data Governance needs to be embedded into the data development cycle and be a part of the CI/CD framework.

The following subsections represent Data Governance-focused capabilities in the Create phase of data.

The Create phase – Data Privacy, Protection, and Security

This pillar must address *data security*. As data is developed, an important consideration is to ensure there is knowledge of any **Personally Identifiable Information (PII)** or sensitive/confidential data to ensure it is properly handled and used. This data needs to be classified with appropriate tags to enable data privacy policies to kick in accordingly. As an example, PII classified data can have a local data privacy policy specific to a country or geographic region that enables tokenization or data masking during data access (a data location policy). Another example is confidentially classified data with a data security policy that only makes data accessible to security officials on a need-to-know basis. There is a distinction between data privacy versus data security. Data privacy is focused on the protection of data based on data privacy policies such as the GDPR. Data security is focused on the authorized access of data. While the actual tagging and classification of data is a part of the Create phase, data enforcement methods are part of the Integrate phase.

Some of the common data classifications are as follows:

- Public
- Restricted
- Confidential
- Sensitive
- Internal use

The Create phase – Data Quality

Quality control needs to be a part of testing during the development of data. Data validation checks and rules need to ensure the integrity and reliability of data. This includes data dependencies and any sequence of tasks during data creation. Data profiling provides insight into the shape of data. Once data is created, there needs to be a conscious effort to monitor the data and provide the results on its level of quality both at rest and in flight. Data observability plays a role here. The results of Data Quality analysis and profiling represent metadata that needs to be made available and associated with a data asset. This step is where a data role, such as a data steward or data product owner, accountable for Data Quality within a particular business domain can define rules and data validation logic by working with a development role to define them.

Data content quality checks need to be in place for the following:

- Accuracy
- Completeness
- Validity
- Uniqueness
- Timeliness
- Data bias

The Create phase – Metadata Management, Data Lineage, and MDM

The three Data Governance pillars must address the *Data is Shared*, *Common Vocabulary and Data Definition*, and *Interoperability* principles. A metadata-driven architecture actively collects and creates metadata at every data life cycle phase. The registration of data enables its discovery. It captures its description, how the business refers to it, its classification, and the schema details to support data interoperability. Master data, such as customer, product, and employee data, should be tagged appropriately. Data sharing guidelines such as **Service-Level Agreements** (**SLAs**) and *Terms of Use* should be defined and made available as part of a data contract to support downstream usage. This includes data retention policies.

This first point of entry (or originating source) in a Data Lineage flow needs to be captured in this phase. This should also include authoritative sources of data, which will be very important in the Integrate phase. This is part of the shift-left mentality, where metadata is collected when the data is developed and created. Waiting until later phases is too late in the life cycle. It creates vulnerability in the discovery and use of data when metadata is captured too late in the process. Collecting schema, formats, business terms, social metadata tags, or user ratings during the Create phase establishes the knowledge necessary for the data to later trigger data policy enforcement, such as Data Privacy and Data Security policies. Not all metadata will be available, but the point is to start with the basic metadata by registering it with name and schema details.

Active metadata starts building metadata for each relationship associated with data. The latest and most up-to-date metadata populated must always be collected and stored. Interested parties must be notified immediately and updated on metadata.

The Ingest phase

Data ingestion is concerned with the movement of data from a source location to a target destination point. The source can be owned by another business domain or belong to the same business domain. However, the goals are the same, ingesting data from one source to load into a target system. Depending on the data source type, you may need to design the data model upfront.

The Ingest phase – Data Privacy, Protection, and Security

Prior to ingesting data, data sharing guidelines associated with the source data in scope must be known. A data contract that depicts the terms of use should be in place prior to designing and implementing a data ingestion flow. This includes the Data Privacy, Protection, and Security policies in a data contract. Data contracts should be made available to data consumers for transparency on the expected rules to follow for data usage, as well as Service Level Agreements.

The Ingest phase – Data Quality

The same criteria as in the Create phase for Data Quality rule generation and data profiling activities apply here. They apply even more in this case, as there is typically little control over what has been defined and put in place by a source system if they are owned by another business domain. The goal is to ensure you are ingesting high-quality data. If not, data anomalies must be resolved to satisfy the business requirements. This can be identified upfront for data at rest in the source system or for data in flight by leveraging data observability.

The Ingest phase – Metadata Management, Data Lineage, and MDM

Leveraging the active metadata component for data discovery allows us to understand data and data relationships. The Data Lineage (for defining the target system) and other operational data, such as processing details, should be captured. Additional business metadata not already collected in the Create phase should be collected in the Ingest phase. This is especially important in the Integrate phase if and when data deduplication efforts need to be completed, which is almost always the case in MDM. Data retention policies need to be understood by data consumers in order to ensure they are enforced later as part of the Archive and Destroy phase.

Next, let's look at the Integrate phase.

The Integrate phase

The integrate phase focuses on applying data transformations that involve integration, aggregation, cleansing, **Extract, Transform, and Load** (ETL) and **Extract, Load, and Transform** (ELT) processes, and data policy enforcement. In this phase, key considerations need to be taken. Data Integration can take place within a source system across several schemas or across distributed data storage in hybrid and multi-cloud environments. This is where great complexity comes into the picture when trying to secure the right level of Data Governance.

The Integrate phase – Data Privacy, Protection, and Security

The reliance on the proper classification of data for each source system is important. Data policy enforcement leverages encryption techniques such as data obfuscation, anonymization, tokenization, and masking, which all transform data. This should handle both data at rest and data in flight. As data is integrated, it's important to be conscious of masked data that violates Data Privacy or Security policies. A well-known fact is that when pieces of data are integrated, they may reveal details that shouldn't be shared, such as PII or confidential data, even if the data has been masked.

The Integrate phase – Data Quality

Data Quality rules and thresholds need to be established for integrated data to validate if the data is fit for purpose. Data defects that have been identified should be addressed in the Integrate phase as part of data cleansing. This involves applying transformations to the data to either de-duplicate, reconcile, or aggregate it to address incompleteness issues or reflect accuracy. The same Data Quality dimensions in the Create phase are applicable here (accuracy, completeness, validity, uniqueness, timeliness, and data bias).

The Integrate phase – Metadata Management and Data Lineage

The rules of thumb applicable in earlier phases apply here, such as the collection of technical, business, operational, and social metadata associated to data. The evolution process of assets into Data Products may start here. Defining a Data Product, its supported use cases, value proposition, and Data Lineage should be inclusive of all assets that make up the Data Product. Other metadata, such as the data contracts associated with a Data Product, can also be captured here. Data interoperability standards need to be in place to enable data discovery and consumption.

The Integrate phase – MDM

MDM plays a significant role in the Integrate phase. MDM is focused on reconciling multiple master data entities across distributed systems and schemas to provide one view of a master entity, such as a customer's. This involves understanding the key identifiers that represent this entity in each system to create an appropriate mapping of data, following established business rules to enable synthesis. The active metadata component helps put the pieces together to enable MDM and improve the overall Data Quality. Extensive data cleansing typically takes place to achieve MDM.

Next to last is the Consume phase.

The Consume phase

The Consume phase is focused on data access and data usage. All the Data Fabric principles in a Data Fabric architecture apply to this phase. It's where the rubber meets the road. This is the most significant phase, as all the previous phases of data to which Data Governance was applied were for proactively governing data to enable a data consumer to derive value from the data during consumption. Data consumer personas are introduced as part of this phase.

The Consume phase – Data Privacy, Protection, and Security

The enforcement of Data Privacy, Protection, and Security rules takes place in the data consumption phase. Data access and consumption must trigger data access management processes, driven by data policies defined to protect data and prevent unauthorized use. This either takes place dynamically when a consumer accesses the data in a storage system or the data has already been protected at rest during storage.

The Consume phase – Data Quality

The expectation here is that high-quality, trusted, fit-for-purpose data has been delivered during consumption. Any data defects should have been identified, cleansed, and resolved in earlier phases of the data life cycle before impacting a data consumer.

The Consume phase – Metadata Management, Data Lineage, and MDM

An audit trail of data usage and consumption with different modes of access should be captured here. This is required to satisfy compliance with Data Privacy regulations such as the GDPR. Data contracts that contain SLAs and terms of use should be available as metadata to data consumers and serve as guardrails during consumption. Social metadata on data usage that captures trends or other similar information should be a part of this phase. Data that is composed/customized for creating new Data Products should preserve Data Lineage.

The last phase is Archive and Destroy.

The Archive and Destroy phase

This phase represents the end of life for data. The data to hand has served its purpose and has stepped through the Create, Ingest, Integrate, and Consume phases. At this point, it must either be archived for later reference to satisfy compliance regulations or permanently deleted.

Data retention policies must be enforced here by data consumers. Data must be either archived or deleted based on the data retention policies. For example, some policies require data to only be retained for a certain time. Data classification is relied upon at this stage to drive automation and easily identify the data that requires applicable data retention policies.

Let's summarize what we've learned in this chapter.

Summary

Data Governance must be designed end to end across all the phases (Create, Ingest, Integrate, Consume, Archive, and Destroy) in a data life cycle. In this chapter, we defined a Data Governance architecture as being metadata-driven and event-driven We also defined what it means to manage metadata as a service, which focuses on the collection, integration, and storage of distributed metadata to derive insights and take action. We learned that a Data Governance layer consists of two components: active metadata and life cycle governance. We also understand that the brain of the Data Fabric architecture is active metadata, which is represented by a Metadata Knowledge Graph. The life cycle governance component consists of capabilities that are centered around Data Governance pillars such as Data Privacy, Protection, and Security, Data Quality, Data Lineage, MDM, and Metadata Management.

In the next chapter, we will focus on the two other layers in a Data Fabric architecture: Data Integration and Self-Service. Data Integration and Self-Service rely on the Data Governance layer. All three layers together represent a Data Fabric architecture.

8

Designing Data Integration and Self-Service

Data Integration and Self-Service represent two sides of the same coin in a Data Fabric architecture. **Data Integration** handles the technical development and processing of data, while **Self-Service** democratizes data. A **DataOps** framework overlays Data Integration and Self-Service to achieve the development and delivery of data with high productivity, speed, and quality control. A Data Fabric architecture comprises both the Data Integration and Self-Service layers and relies on the Data Governance layer for embedded and automated governance.

In this chapter, we'll introduce you to the Data Integration and Self-Service layers of a Data Fabric architecture. We will discuss how these two layers interact with one another and how they work with the Data Governance layer to create a Data Fabric. The applicability of Data Fabric and DataOps principles will be discussed as part of the life cycle of data. We will conclude the chapter by presenting a Data Fabric reference architecture consisting of three layers: Data Integration, Data Governance, and Self-Service.

In this chapter, we'll cover the following topics:

- DataOps-based architecture
- Data Integration layer
- Self-Service layer
- Data journey
- Data Fabric reference architecture

DataOps-based architecture

DataOps is a mature data management framework with best practices and principles adopted from software engineering and manufacturing frameworks such as Agile, DevOps, and statistical process control. In DataOps, these successful best practices are applied in the context of data management to achieve fast, high-quality, and cost-efficient data delivery. DataOps continuously monitors data post-deployment to keep an eye on its pulse to check its health. In my view, a DataOps discipline is a necessary framework that drastically accelerates a Data Fabric architecture. DataOps was introduced in *Chapter 4, Introducing DataOps*.

> **What's the difference between DataOps and data engineering?**
>
> These terms are often used interchangeably; however, they are not the same thing. The assumption made when referring to DataOps is that it refers to daily data operations within a business. DataOps is a data management discipline with established principles that achieve agility, automation, speed, and quality control. DataOps can be used in data engineering to deploy quality data assets and Data Products and to proactively monitor its health. However, executing data engineering doesn't mean you are executing a DataOps framework.

Figure 8.1 is a refresher on the DataOps principles introduced in *Chapter 4*. These principles are embedded in the Data Integration and Self-Service layers of a Data Fabric.

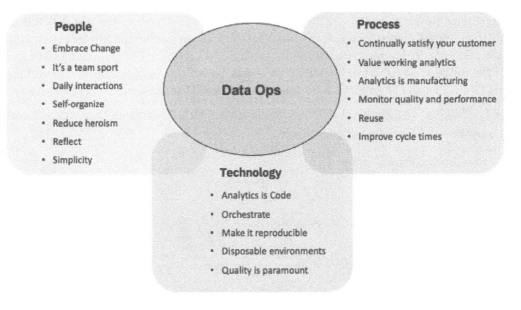

Figure 8.1 – DataOps principles by people, process, and technology

A Data Fabric architecture incorporates DataOps principles to develop and deploy data throughout its life cycle. These principles serve as guideposts to ensure the architecture properly supports the processing and delivery of data maturely and efficiently. In *Chapter 7*, we reviewed the Data Governance layer in a Data Fabric architecture. We established that the foundational architecture patterns are metadata-driven and event-driven. There are other architecture patterns in a Data Fabric architecture, such as microservices, APIs, and more. However, our emphasis is on discussing the key data architecture concepts that distinguish a Data Fabric architecture.

Let's review the Data Integration layer of a Data Fabric architecture.

Data Integration layer

The Data Integration layer of a Data Fabric architecture represents a set of capabilities in the form of services, tools, and data environments (development, test, and production) that support the development cycle of data to achieve downstream consumption. The Data Integration layer is composed of two pillars: **data management** and **development workflow**. Each pillar represents a group of capabilities and technologies that focus on realizing a phase in the life cycle of data. The end goal of the Data Integration layer is to technically support the ingestion, integration, and delivery of data. It needs to support the necessary provisioning and configuration to enable data to be consumed.

> **How does a Data Fabric support data assets, data-related assets, and Data Products?**
>
> Data can be a **data asset** or a **data-related asset** that can be grouped to represent one of the several assets in a **Data Product**. What differs between a data asset, a data-related asset, and a Data Product is the business and governance guardrails it needs to follow. Several definitions in the industry have slightly different perspectives. I offer a definition in *Chapter 7, Designing Data Governance*. Here is a summarized version: data assets represent enterprise data that is curated, cataloged, and should be governed. It can represent raw data. Data assets aren't always created with the intention of sharing at a high scale. Data Products are intentionally created to enable high-scale data sharing, have a specific business value proposition, and can contain several data assets and data-related assets. A Data Fabric architecture supports all three concepts.

In the next section, let's start with the first pillar in the Data Integration layer: data management.

Data management

The data management pillar consists of three components: data ingestion, queries, and data transformation. Data ingestion supports a multitude of different types of source systems to enable the movement/ingestion of data. Diversity is key, and most important is the handling of metadata for each source system. Metadata generated during data ingestion must be collected and integrated as part of the Data Governance layer in a Data Fabric architecture. While a Data Fabric architecture supports the ingestion or movement of data into a centralized data system, it's flexible in supporting diverse types of source systems and data management models.

What differentiates a Data Fabric architecture is its active metadata-driven approach and embedded Data Governance. So when it comes to data management, Data Fabric represents an abstract layer that sits on top of the physical data layer. It's capable of representing a monolithic data storage system or distributed data storage systems. I emphasize this point as there are views in the industry that Data Fabric is a centralized storage architecture, which is not the case from my point of view. A Data Fabric architecture is driven by the needs and direction of the business architecture. In *Chapter 6*, we discussed the four kinds of architectures that establish an enterprise architecture:

- Business architecture
- Data architecture
- Application/system architecture
- Technology architecture

Data Fabric architecture can start by supporting a centralized data system, such as a data warehouse, and evolve to add more systems, such as data lakes and data lakehouses.

The following is a base list of data ingestion patterns in a Data Integration layer:

- Data connectors
- Change data capture
- **Application programming interfaces (APIs)**
- Streaming

The data transformation component in the data management pillar is focused on data processing, such as taking data in its original form and changing it by aggregating it, replacing it, cleansing it, and more. This can take place in batch or streaming form. The query component represents technologies that enable the querying and reading of data. Distributed queries that integrate data across source systems and create a view represent a form of Data Integration. Data Lineage plays an important role in the provenance of data from its original state to its derived state, whether transformed or queried. The objectives of a Data Fabric architecture are to support several technologies and approaches to data processing and pipeline development. The goal is to do so alongside the generation and collection of metadata to govern and automate data management activities. This is where the Data Governance layer plays a role with the Data Integration layer.

These are the data transformation and query styles in a Data Fabric architecture:

- **Extract, Transform, and Load (ETL)** or **Extract, Load, and Transform (ELT)**
- Data as code or a programmatic approach
- Data virtualization/distributed queries
- Federated queries

- Ad hoc queries

- Streaming

Next, let's discuss the development workflow pillar in the Data Integration layer.

Development workflow

The **development workflow** pillar handles the end-to-end development cycle of data. These capabilities execute on DataOps principles and support development across five stages: develop, orchestrate, test, deploy, and monitor. You will find similarities between terms used in DataOps and those used in DevOps, such as **automated testing** and **CI/CD**. However, DataOps focuses on a subset of DevOps, which is data management.

The development workflow pillar has the following areas:

- **Data orchestration** – Automated orchestration of data, tools, and infrastructure that support data ingestion and data transformation activities.

- **Data and code version control** – Version control systems that keep track of changes in code, data, and configurations.

- **Automated testing** – Automation is key here for code, data, configurations, and so on.

- **Continuous Integration/Continuous Delivery and Deployment** – Applicable to data and its configurations.

- **Data environments** – Support the integration and deployment of Data Products across environments (development, test, and production).

- **Data monitoring** – Ongoing monitoring of data after deployment. This directly supports having a pulse on the health of data.

In the next section, we will discuss the third layer of a Data Fabric architecture: Self-Service.

Self-Service layer

The Self-Service layer's main objective is to democratize data. It enables the exploration and use of high-value data. The Self-Service layer continues from where the Data Integration layer left off. A Self-Service architecture has a focused role in enabling the sharing and consumption of data via Self-Service processes. It works with the Data Governance layer to govern and protect data while unlocking its use across diverse systems. Metadata- and event-driven architectures introduced in the Data Governance layer are relied upon. The Self-Service layer also relies on the Data Integration layer to facilitate, via automated means, the delivery of data to data consumers. Data consumption is typically read-only, however, exceptions are possible and need to be designed into the architecture. An example is the need for special privileges for superusers.

There is also the notion of enabling data consumers via Self-Service data composability capabilities to customize data for their needs. Data Fabric principles (*data are assets that can evolve into data products, data is shared, data security*, and *data is accessible and interoperable*) take an active role in the Self-Service layer.

The Self-Service layer in a Data Fabric architecture has two pillars: data democratization and data consumption. Let's review the data democratization pillar.

Data democratization

The data democratization pillar handles data registration, management, and monetization. It focuses on creating reusable Data Assets and Data Products. It caters to data sharing, both internal and external. Internal data sharing is constrained within the boundaries of an organization, and external data sharing is across different companies and vendors. Data monetization supports the selling and buying of data, which is typically associated with subscriptions. It can also represent internal data monetization concepts that require the tracking and usage of data for internal financial recovery. Data Privacy, Protection, and Security play a major role in the Self-Service layer of a Data Fabric architecture.

The following components are a part of the democratization pillar:

- **Data registry** – The data registration process in the Self-Service layer establishes the existence of data assets, data-related assets, and Data Products. It enables data to announce itself and make itself discoverable by data consumers. This entails cataloging and publication mechanisms for data with contextual metadata that properly identifies and defines data objectives.

- **Data contract** – A data contract establishes certain levels of guarantees it will offer data consumers such as **service-level objectives/agreements (SLOs/SLAs)**. This should be in line with the data delivery methods. A data contract also addresses the terms and conditions, or the rules and policies a data consumer agrees to comply with.

- **Data monetization** – Manages metered data such as pricing, entitlement, subscriptions, and data usage. Enables consumers to automatically receive data based on the selected mode of delivery. Data monetization can represent pricing of data but can also represent other aspects of monetization for internal use such as adoption via high usage.

- **Data delivery methods** – Manages consumption methods associated with entitlements to enable delivery and access to data.

- **Data product management** – Supports active data product management across a data life cycle that includes operationalization activities to achieve evolving data as an asset to a data product.

- **Data health** – Provides the overall health of data to enable quality improvements. This includes usage rates, defects, and ratings. It establishes a feedback loop from data consumers that can be acted upon by data producers.

Let's review the data consumption pillar in the Self-Service layer.

Data consumption

This is where everything comes to fruition, when a data consumer is able to use the data. After data has been registered and enabled for sharing via the data democratization pillar, capabilities that enable its discovery and consumption are available via the data consumption pillar. The goal is to enable data consumers to make data-driven decisions with the intent of taking action leading to profitable and cost-efficient decisions.

The following components are a part of the data consumption pillar:

- **Data exploration and discovery** – Data consumers can explore, search, find, and understand available data in an enterprise.
- **Data access and delivery management** – Data access that includes approvals, authorization, provision, and delivery.
- **Data composability** – Composing of consumed data as part of the data consumption process. Enables customization of data.
- **Feedback loop** – Communication channel between data consumers and data producers to achieve ongoing quality improvements.

In the next section, we will step through a day in the life of data in the Data Integration and Self-Service layers of a Data Fabric architecture.

Data journey in a Data Fabric architecture

The best way to understand of how a Data Fabric architecture handles data is to look at it from a journey standpoint. We will follow the same approach as we did in explaining the Data Governance layer in *Chapter 7*. Technical capabilities within a Data Fabric architecture enable the development of data and its delivery. DataOps principles act as a medium in the deployment of data within the Data Integration and Self-Service layers of a Data Fabric architecture. DataOps focuses on efficiencies and quality control in the development, deployment, and delivery of data.

Data travels through five logical phases in a data life cycle across the Data Fabric architecture:

1. Create
2. Ingest
3. Integrate
4. Consume
5. Archive and destroy

DataOps has its own cycle with specific activities that guide the development of data from the *create* to the *consume* phase. The stages in DataOps consist of development, orchestration, testing, deployment, and monitoring. At the end of the DataOps *deploy* stage, data has been deployed into a production environment and is ready for consumption by the business. *Figure 8.2* overlays the five data life cycle phases with the DataOps stages within a Data Fabric architecture.

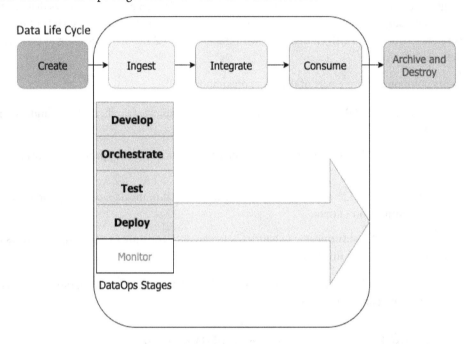

Figure 8.2 – Data life cycle phases to DataOps stages

Data can step through many iterations during the development cycle. Data defects identified during the DataOps *test* stage will require further development or orchestration to manage the changes. There isn't a strict sequence on how the stages apply but rather, what is important is that DataOps principles are applied at every DataOps stage applicable in the data life cycle.

In the following five sections, we'll step through the data life cycle phases. We'll review how the Data Integration and Self-Service layers manage data with DataOps, and embedded governance via the Data Governance layer.

Phase 1 – Create phase in the Data Integration layer

Data can be created via a multitude of methods. The assumption made in the *create* phase is that the development and deployment activities took place outside of the Data Fabric's Data Integration layer and are sitting in a storage system. Some examples could include a business application that captured a customer order and stored it in an **Online Transactional Processing (OLTP)** system or an **Internet of Things (IoT)** device that collected and sent data to storage. In such a case, there is a need to either ingest or integrate data from a storage system(s) leveraging the Data Integration layer.

Phases 2 and 3 – Ingest and Integrate phases in the Data Integration layer

Data ingestion is the second phase in a data life cycle but the first phase that steps through a Data Fabric architecture. For the purpose of explaining data ingestion, the assumption made is that the source data is owned by another team where there is no control over their data architecture approach. However, there could be cases where data ingestion from a source to a target is managed by the entire Data Fabric architecture. Physically, data is stored in a data storage system(s) as part of the *create* phase. Data ingestion into a target system(s) can be achieved with several styles such as ETL, ELT, streaming, and Data-as-Code. Data ingestion isn't always necessary when data persistence isn't required. A read-only approach could be taken leveraging integration styles such as federated SQL queries or data virtualization technology. Most data will eventually have some transformation applied, whether in the form of cleansing, enrichment, masking, or other data preparation across its journey. However, raw data can be leveraged without any applied transformations.

The data registry process supports the *ingest* phase as that is the first time there is an acknowledgment of new data within the Data Fabric architecture (this is managed in the Self-Service layer, which is discussed later). Ideally, metadata has been collected in the *create* phase that could be leveraged as part of the *ingest* phase. If this is not the case, then a Data Fabric architecture needs to be capable of accounting for both cases when metadata is present and when it's not. Either way, even if metadata is present, it may be of low quality; therefore, a Data Fabric architecture must address this via an automated metadata enrichment process. The tagging, classification, and profiling of data is a prerequisite that must be achieved for all data flowing through the Data Fabric architecture. Metadata populated during data ingestion and orchestration flows, as well as integration pipelines, represents operational metadata. This includes database logs, job runs, and Data Lineage details. Metadata collection and integration take place in the Data Governance layer; therefore, metadata populated within the Data Integration layer must be communicated to the Data Governance layer as streaming events.

The *integrate* phase for data is a continuation of the *ingest* phase. It is continuously collecting metadata (technical, business, operational, and social metadata) generated in the Data Integration layer and establishing the necessary relationships and data knowledge. Data Quality, Data Lineage, Master Data Management, Data Privacy, Protection, and Security measures are handled by the Data Governance layer. Another consideration here is that during data ingestion and Data Integration, it's important to validate adhering to Data Governance standards and guidelines. This can range from specific formats established for managing data, to how the data should be used, and more. This is where relying on the Data Governance layer is applicable.

In the next few sections, we will discuss data ingestion and integration across the five DataOps stages.

Ingest and Integrate phases – DataOps develop stage

DataOps principles: embrace change, it's a team sport, daily interactions, self-organize, reduce heroism, simplicity, analytics is code, and reuse.

The *develop* stage represents the start of the data development cycle, which includes data ingestion and integration pipelines. Creating a culture and business environment that manages data efficiently should start with a collaborative and efficient culture across data producers roles (business and data) involved in the development of data, and being a team player where constant communication is executed. Technical data roles, such as data engineers, developers, and data scientists, stay connected on the dependencies in the pipeline they own and manage, while they work side by side with the business to ensure data developed meets the business requirements and respects governance guardrails.

DataOps capabilities execute on the *analytics is code* principle, which offers a high level of automation. It includes continuous integration and **Infrastructure as Code (IaC)** approaches. Version control features enable the rollback of changes and establish an audit trail. Data storage can be set up programmatically by leveraging IaC. The value statement here is like that of DevOps: managing data as code provides more control and customization, which enables high automation. A programmatic approach to data management can and should work with low-code/no-code tools. This is an important factor to highlight as the premise here is that both approaches should be leveraged in a complementary manner and supported in a Data Fabric architecture to empower business and technical roles to create value-driven data.

Handling the diversity of data, whether in the form of storage, ingestion and integration technologies, or interfaces, is necessary. It needs to account for the types of source systems, such as OLTP, which support critical business operations and do not support the manipulation of data from external data processes as compared to an **Online Analytical Process (OLAP)** system. An event-driven architecture can play a role here in the loading of new/updated transactional data from an OLTP source to a target OLAP system to enable analytics.

> **Note**
>
> While this book mentions a list of technologies, approaches, and source systems that need to be supported in a Data Integration layer, the actual building of a Data Fabric architecture can be a gradual exercise. It can start small with a selected number of approaches and grow out as needed. Prioritizing the technical data approaches based on the business needs is one way to start.

A Data Fabric architecture can provide a huge value statement and create wonders for the data methods and types it supports, but if it doesn't support what the bulk of the business operates on, its value is minimal.

Ingest and Integrate phases – DataOps orchestrate stage

DataOps principle: orchestrate.

The data orchestration stage applies to batch-based data pipelines. It automates and manages data pipelines based on a specific sequence of tasks. It can manage the ingestion and loading of data across different data storage systems, and take different paths based on conditional criteria being met or not. It typically uses a **Directed Acyclic Graph (DAG)**, which is a graph that represents a series of activities executed in a directed manner via edges and nodes. Automated data orchestration is a key DataOps principle. An example of orchestration can take ETL jobs and a Python script to ingest and transform data based on a specific sequence from different source systems. It can handle the versioning of data to avoid breaking existing data consumption pipelines already in place. The authors, *Joe Reis* and *Matt Housley*, in their book *Fundamentals of Data Engineering*, suggest the use of data orchestration to coordinate several development cycles and to manage different data versions. This is an important trait in Self-Service data sharing.

Data orchestration offers a superior approach that focuses on automation as opposed to traditional approaches that may create piecemeal data jobs without any coordination. Data orchestration across multiple data environments (on-premises, single cloud, or multi-cloud) can take place. Automated data orchestration can be leveraged in different settings. Data orchestration can happen at different levels of granularity that meet the required data architecture or data management style. Data orchestration can be contained in the development of a specific Data Product that consists of several assets, or it can represent a broader set of tasks that span multiple assets across different Data Products.

Ingest and Integrate phases – DataOps test stage

DataOps principles: quality is paramount, analytics is manufacturing, and make it reproducible.

Testing is centered around two things: the quality of data and the code that manages that data. The goals are to produce high-quality data and satisfy performance by identifying defects during the development cycle by leveraging automated testing and defect remediation. It's important to be able to reproduce data defects to drive resolution. Issues identified in the monitoring stage that were resolved can be tested for validation in this stage. Quality control is a foundational staple in a DataOps framework and takes place both in the development cycle and post-production of data to achieve quality. Measuring the quality of data will differ depending on batch versus streaming-based methods. Quality measurement for streaming data is a lot harder and must be planned more carefully due to the nature of the potential incompleteness of data presented in real time.

Ingest and Integrate phases – DataOps deploy stage

DataOps principles: value working analytics, continually satisfy your customer, and improve cycle times.

Continuous delivery and deployment of data takes place in this stage. Data must be accurate, reliable, and trustworthy to provide value to a business. Success in this stage is represented by the satisfaction of the customer (data consumers). A constant customer feedback loop during development aims to ensure the success of deployed data. This also provides an opportunity to evaluate the development cycle to determine any improvements that could be implemented in order to deliver Data Products more quickly with higher quality. Once data is deployed and made available for use by data consumers, it moves into the *consume* stage.

Ingest and Integrate phases – DataOps monitor stage

DataOps principle: monitor quality and performance.

The *monitor* stage has a strong foot in the Data Quality pillar of the Data Governance layer. Data monitoring is the ongoing monitoring of Data Quality and reliability. This stage is focused on identifying any potential defects, whether performance or data integrity-related, that impact the ability to access and use data. Designing the ability to receive automated notifications and alerts for data sources/interfaces that are failing would enable a data practitioner to handle them accordingly. A data consumer can choose to delay the use of data they rely on until the issues are resolved. Monitoring can also focus on validating whether a data source is complying with an SLA. This can enable data roles such as data engineers and developers to take proactive measures to avoid breaking data pipelines.

It serves as a control point for potential defects in production-level data pipelines. Data monitoring identifies issues such as data drift, source outages, data security, data access, and Data Quality. Overall, it should understand the health of data. From a data engineering angle, it should monitor uptime, latency, and data volume (as discussed by Joe Reis and Matt Housley in *Fundamentals of Data Engineering*). Any potential issue impacting the business should be designed as a checkpoint that is monitored in this phase. Data monitoring can be applied at two varying degrees: it can be thought of as dependencies on source systems (outside of the Data Fabric architecture) or deployed data (within the scope of the Data Fabric architecture) to enable consumption.

In the next section, we will review the *consume* phase in the Self-Service layer.

Phase 4 – Consume phase in the Self-Service layer

Data consumption starts by identifying data as shareable across an organization. The Self-Service layer of a Data Fabric architecture supports the notion of data assets, data-related assets, and Data Products. The premise is that data needs to be discoverable and understood in order to provide an inventory of what is available for use by a data consumer. The data registry component enables data to be announced via the collection of metadata. It supports data producers in the process of design and development. Data assets evolve into Data Products with active product management capabilities applied.

Data product life cycle management guides data producers in the operationalization of Data Products, such as in the creation of a data contract and delivery methods to enable high-scale data sharing. Data monetization capabilities enable the purchasing of data or tracking data usage. This includes managing pricing, entitlements, and subscriptions. The Data Governance layer supports the Self-Service layer in the democratization of data that is secure, protected, and governed. During data consumption, enforcement of Data Privacy, Protection, and Security policies are executed. Data Lineage tracks data access and delivery.

Data delivery within a Data Fabric architecture can serve as input to other pipelines, such as in MLOps for data science use cases. The Self-Service layer supports a diverse set of business intelligence and analytical use cases that expect and rely on high-quality data. Let's review how the Self-Service layer operates in a data science use case that relies on data within a Data Fabric architecture to build an ML model.

There are three stages in an MLOps workflow, as discussed by *Emmanuel Raj* in *Engineering MLOps*, Packt Publishing:

1. Build

2. Deploy

3. Monitor

Let's look at each stage in detail, starting with the *build* stage.

MLOps build stage

The MLOps build stage represents the model's development cycle. The first step in the build stage is to get trusted, quality, and secure data quickly. This data is created in a Data Fabric's Data Integration layer, established for sharing in the Self-Service layer with embedded governance via the Data Governance layer. The data registry component leverages metadata capabilities to enable data discovery. A data scientist will engage with the Self-Service layer via its data exploration and discovery features. A data contract establishes the terms of data use and available SLAs. Data access management identifies and authorizes access to data based on data policies managed in the Data Governance layer. Data delivery is coordinated by the Self-Service layer and Data Integration layers with high efficiency and speed.

MLOps deploy stage

Testing of the model across different spectrums is executed here. Data defects that surface are reported via a feedback loop component. Data consumers have a feedback loop for communicating issues and providing feedback for quality improvements. Data Lineage kicks in to capture the evolution of data leveraged to create an ML model.

MLOps monitor stage

Data monitoring capabilities within a Data Fabric architecture are leveraged to proactively identify issues.

Phase 5 – Archive and Destroy phase

There are a number of different factors that contribute to the decision of moving data into an archive and destroy phase. It could be based on established data regulations that require data to be stored for a short period of time. Another can be that data no longer provides value and hasn't been used for an extended period of time. The combination of capabilities across a Data Fabric architecture, such as data monitoring, data health, data contract, and governance data policies, would determine the right course of action.

In the last section of this chapter, we will put it all together and highlight the key points across all three layers (Data Integration, Data Governance, and Self-Service) in a Data Fabric architecture.

Data Fabric reference architecture

We have introduced and reviewed the design of three layers in a Data Fabric architecture. This establishes a Data Fabric reference architecture that could be leveraged as an approach in a digital transformation journey. It can also serve as a reference to build a roadmap on what, when, and how you can progress toward the building or buying of solutions to build a Data Fabric architecture.

Data Fabric architecture highlights

Let's briefly highlight key points in each of the three layers of a Data Fabric architecture.

Data Governance layer

This represents the checks and balances of a Data Fabric architecture. It contains the brain (Metadata Knowledge Graph) that powers the body (Data Integration and Self-Service layers) to support the development and delivery of data that is quality controlled, trusted, and enforced. The Data Governance layer heavily relies on a sophisticated, active metadata- and event-driven approach. It applies Data Quality, Data Privacy, Protection and Security, Data Lineage, and Master Data Management to govern data across all data life cycle phases.

Data Integration layer

This manages the development, deployment, monitoring, and delivery of data. It does this respecting critical DataOps principles and Data Fabric principles incorporated in the Data Fabric architecture. Data ingestion, data transformation, queries, and data pipelines are all aspects of data processing activities in the Data Integration layer. Data Integration depends on the Data Governance layer and interacts with the Self-Service layer.

Self-Service layer

Supports data consumers in Self-Service access of high-quality, rapid, and trusted data. It empowers data producers to share data. It enables the operationalization of data via data contracts, data monetization, subscriptions, and multiple delivery modes. The feedback loop establishes a path for data consumers to provide feedback to data producers, leading to quality improvements. Data composability offers Self-Service data customization. All capabilities in the Self-Service layer depend on the Data Governance layer to enable data that is secure, protected, and trusted. It also interacts with the Data Integration layer to implement the technical execution that achieves data access and delivery.

Figure 8.3 demonstrates a Data Fabric reference architecture:

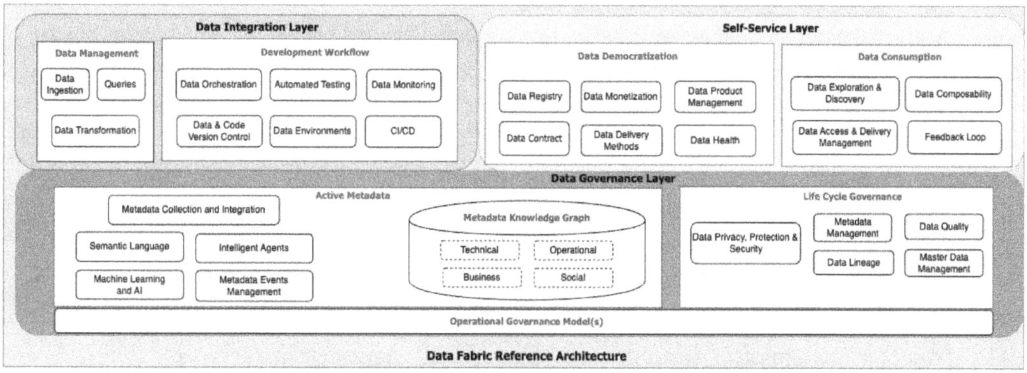

Figure 8.3 – Data Fabric reference architecture

The Data Fabric reference architecture illustrated is not the only way to design a Data Fabric architecture, nor is it the ultimate or definitive solution for all Data Fabric requirements. There can be variations that represent this architecture tailored to the business goals and drivers of an organization. The capabilities represented in each layer provide a reference on what to target, but the journey can start with a subset. The key is to ensure you have a solid foundation in the Data Governance layer of a Data Fabric architecture that will create the robustness to support a composable architecture.

> **Tip**
>
> Use the Data Fabric reference architecture to build your target vision or compare and contrast with an existing data architecture. This can help establish a roadmap of where you aspire to get to. It can also be used as a gap assessment to determine where you are and what is missing.

Leverage the reference architecture to help guide the assessment of tools and technologies. It also serves as a position statement to understand what represents a Data Fabric architecture. You may agree or disagree and have a proposal for further refinements driven by the base established here. I invite them and welcome you to pursue them.

Summary

A Data Fabric architecture manages, governs, and supports the life cycle of data. It embraces a shift-left approach by infusing metadata and event-driven architectures with foundational Data Governance and data sharing capabilities. In this chapter, we reviewed the Data Integration layer focused on the development of data with a DataOps lens across the stages (develop, orchestrate, test, deploy, and monitor). We also reviewed the Self-Service layer focused on the democratization of data to achieve data sharing. Both the Data Integration and the Self-Service layers are of extreme importance; however, they heavily rely on embedded and automated governance in the Data Governance layer. All three layers seamlessly work together to manage, govern, and share data with maximum business value. They represent the constructs that compose a Data Fabric architecture.

In the next chapter, we will review how to design a technical architecture for a Data Fabric.

9

Realizing a Data Fabric Technical Architecture

An implementation of a Data Fabric architecture consists of an ecosystem of different tools, data systems, and technologies. Beyond that, it requires the right set of capabilities and data processes that support Data Fabric principles. All these pieces work together as a cohesive automated ecosystem driven by a Data Fabric architecture design to create a connected fabric of data. There isn't one strict path toward the implementation of a Data Fabric architecture. There are several variations of how you might choose to implement it guided by business goals and strategy. What's important is to leverage the design guardrails discussed in this book to set you on the path toward the realization of a successful Data Fabric technical architecture.

In this chapter, we will cover the technical architecture of a Data Fabric and what tools, technologies, and capabilities are necessary and should be considered for what reasons. A walk-through on the technology that shapes each of the three layers in a Data Fabric architecture will be discussed: Data Governance, Data Integration, and Self-Service. Two use cases will be reviewed, distributed data management and regulatory compliance, to showcase how a Data Fabric addresses the requirements of each layer. By the end of this chapter, you will understand the kinds of tools and components needed in the implementation of a Data Fabric architecture. You will also have the perspective of a reference architecture that applies to both a Data Fabric and Data Mesh architecture.

In this chapter, we'll cover the following topics:

- Technical Data Fabric architecture
- Use cases
- Data Mesh multi-plane requirements
- Data Fabric with Data Mesh reference architecture

Technical Data Fabric architecture

A Data Fabric platform is infused with intelligent metadata that acts as a parallel universe, tracing and understanding the flow of data across its various stages, tools, systems, and processes. You can think about it as the kids' game "copy me," where one copies everything another says or does. In the case of metadata, it works in a similar manner except that it also derives insights and knowledge as the metadata is collected. It learns from its navigation and proposes improved approaches to data management. Active metadata is at the center of the Data Fabric universe. It focuses on enabling data that is fit for purpose, secure, and delivered with speed while increasing efficiency. Active metadata captures, from creation to end of life, all data details such as facts, decisions, relationships, and policies, to then take action by influencing its strategic and efficient direction.

A technical architecture facilitates the implementation of a data and application architecture with technology choices. A Data Fabric's technical architecture is composable and modular and consists of several tools and technologies. The selection of required tools is driven by its capabilities and ability to address Data Fabric principles.

> **Note**
> This chapter won't offer an assessment or recommendations on vendors or other tools. The focus is to describe the *kinds* of tools in the market today that could be leveraged.

When selecting tools, technologies, and other services, weighing the pros against the cons via a trade-off analysis is a must. A trade-off analysis is an architectural approach for reaching architectural decisions (*Software Architecture: The Hard Parts*, by *Neal Ford et al.*). In the next section, let's identify the kinds of tools and technologies needed in each layer of a Data Fabric.

Data Fabric tools

A specific set of capabilities are required in each Data Fabric architecture layer. The corresponding tools that realize these capabilities will vary depending on the vendor, open source, or custom-built services. While I list the *kinds* of tools, I make some assumptions about the capabilities the tools offer. The greater focus should be on the required capabilities listed to realize a Data Fabric architecture. An architect can then evaluate a tool to determine whether the required capabilities are met. As an example, a vendor data catalog may have data curation, asset inventory management, Data Quality, Data Privacy, and Data Lineage capabilities, but another vendor data catalog may have a subset of capabilities such as asset tracking and basic curation capabilities.

What's important to note here is that tools and technologies are evolving every single day. There is no shortage of technologies in the data management space. As such, the list of tools is a point-in-time statement that will need to be revisited often to ensure you can maximize the value of a Data Fabric architecture. As *Joe Reis* accurately points out in his book *Fundamentals of Data Engineering*:

"An architecture is a strategic vision and focus, the tools and technologies that realize the architecture are tactical."

In the following three sections, we will look at the kinds of tools and associated capabilities in each layer of a Data Fabric architecture.

Data Governance layer

The Data Governance layer governs data in the ecosystem of connected fabrics, with active metadata gluing each data endpoint together. The following is a list of possible tools with the required capabilities:

Data Governance Capabilities	Kinds of Tools
<ul><li>Data inventory that captures and enables exploration of data sources, applications, data assets, data-related assets (such as ML models), trusted sources, schema, column, and attributes</li><li>Business groups to categorize assets (business domains and subdomains)</li><li>Business semantics (business terms, synonyms, and data definitions)</li><li>Data classification to enable data protection such as sensitive, confidential, government, and other types of data classifications</li><li>Operational and analytical Data Lineage that establishes provenance by capturing the data journey with the right level of granularity down to the column/attribute level</li><li>Data Quality score and analysis that identifies data anomalies and validates fit-for-purpose data</li><li>Data profiling that enables understanding the shape of data</li><li>Data model management – metamodel mapped to a physical data model:<ul><li>Business semantics to data inventory</li><li>Data classifications associated with data inventory</li><li>Data policies for data classifications</li></ul></li><li>Data policies management (data protection, data privacy, Data Quality, and more)</li><li>AI and ML-infused Metadata Management, such as curation and data classification</li></ul>	<ul><li>Data Governance catalog</li><li>Data Quality tools</li><li>Data science model catalog</li></ul>

Data Governance Capabilities	Kinds of Tools
• Data privacy and data protection policies authoring and enforcement to achieve data masking, data encryption, and data obfuscation • Secure data access leveraging data classification and data policies • Data access control authenticating user/system and granting access at the role level, attribute level, or leverage data classification labels	• Data Governance catalog • Data protection tools • Access management • Security authorization
• Federation and/or sharing of data/metadata across metadata repositories	• Federated metadata synchronizer

• Master Data Management:
   ▪ Entity relationship management and resolution across multiple data sources
   ▪ Integration, aggregation, and cleansing of master data
• Reference data management
	• Master Data Management
	• Reference data management

• Metamodel, data model, associated properties, and relationships management and storage • Metadata Knowledge Graph • Access and maintenance of data models	• Knowledge Graph database • Query language • Data Governance catalog
• Proactive data monitoring for inflight data • Alerts and notifications of defects with sufficient incident details to drive action	• Data observability

Figure 9.1 – Data Fabric's Data Governance layer capabilities and kinds of tools

Let's review the capabilities and kinds of tools in a Data Fabric's Data Integration layer.

Data Integration layer

The Data Integration layer focuses on supporting technical roles in data processing activities such as ingestion, data preparation, and data delivery. This layer supports a variety of technical patterns and tools. The extent of tools required in a Data Integration layer is driven by the technologies that align with the strategy of an organization. As an example, is real-time data processing critical? Is there more emphasis on batch-based data processing? Are there legacy systems from which data needs to be extracted and managed? In the following table, I've listed the capabilities and kinds of tools in a Data Integration layer:

Data Integration Capabilities	Kinds of Tools
• Data pipelines management (batch) • ETL and ELT approaches	Data Integration/transformation
• Programmatic data processing (Manage data as code): • Programming languages such as Python are used to build scripts and pipelines	• Data science/engineering tools such as notebooks • Programming/development tools
• Querying of data	SQL-based or other query tools
• Near real-time data management and processing • Event management	• Streaming technology
• Distributed querying of data across source systems • Data protection policy enforcement	• Data virtualization
• Data replication for delta changes	• Change data capture
• End-to-end data pipeline orchestration • Workflow management	• Data orchestrator
• Data storage	• SQL or NoSQL databases including data warehouses, data lakes, data lakehouses, and graph databases
• Code storage applicable in a development cycle • Developer collaboration to achieve the branching and merging of development, testing, and deployment of code • Version control • Automated testing • Continuous integration, delivery, and deployment • Infrastructure as code • Containers management	• Code repository • Testing tool • CI/CD tools • IT configuration management • Container platform

Figure 9.2 – Data Fabric's Data Integration layer capabilities and kinds of tools

As new data processing technologies emerge and are added to the Data Fabric architecture, it's critical to design its cohesion with the Data Governance layer and Self-Service layer. What distinguishes a Data Integration layer from a typical data engineering architecture is that it inherently designs active metadata, Data Fabric, and DataOps best practices. The Data Integration layer operates jointly with the Data Governance and Self-Service layers, and not independently.

Self-Service layer

The Self-Service layer of a Data Fabric democratizes access and use of data. It supports data owners, Data Product managers, and Data Product owners in the creation and management of data with the intention of realizing Self-Service data consumption. There are many ways in which this layer could be designed from a technology standpoint. The following is a list of capabilities and the kinds of tools that can be used:

Self-Service Capabilities	Kinds of Tools
• Data Product exploration and discovery • Life cycle management of Data Products • Data operationalization • Self-Service data delivery and consumption • Data contracts management • Data monetization • Data composability • Feedback loop	• Data marketplace
• MLOps and AIOps • Model development, testing, and deployment	• Data science/MLOps tools
• Visualize and analyze data • Dashboards and reports	• Business intelligence • Data visualization

Figure 9.3 – Data Fabric's Self-Service layer capabilities and kinds of tools

The key tool in this space is a data marketplace that works cohesively with the Data Governance and Data Integration layers. I've often seen many architecture designs that define a data marketplace as a lightweight portal that sits on top of existing tools and data repositories. While it needs to accomplish this and orchestrate data management activities, it also needs to have a special focus on managing the life cycle of Data Products with the goal of achieving Self-Service data sharing. A data marketplace integrates with Data Governance tools, such as a data catalog, and DataOps tools, such as data preparation, provisioning, and transformation tools. It also integrates with other Self-Service tools, such as data visualization and data science.

We have discussed the kinds of tools and capabilities that comprise a Data Fabric. To support this, there need to be infrastructure components that enable the hosting of these tools and data management via on-premises, hybrid cloud, or multi-cloud environments.

In the next section, we will review a sample list of vendor and open source tools.

Vendor and open source tools

There is a broad range of tools in the market, whether private or open source. The list of possibilities is enormous. I'm not advocating for one tool versus another – that is left up to the architect to decide based on their trade-off analysis. The following is a list of some of the available tools categorized by each Data Fabric layer (Data Governance, Data Integration, and Self-Service). The list is in alphabetical order and isn't exhaustive:

Data Governance Layer	Data Integration Layer	Self-Service Layer
• Apache Atlas • AWS Glue • Azure Purview • DataHub • Egeria • IBM Guardium • IBM Watson Knowledge Catalog • Immuta • MANTA • Neo4j • Ranger	• Apache Airflow • Apache Kafka • Apache Spark • Databricks • Data build tool • DataKitchen • Denodo • GitHub • IBM Databand • IBM DataStage • IBM Watson Query • Jenkins • RabbitMQ	• AWS Marketplace • IBM Watson Studio • Jupiter Notebook • Microsoft Power BI • Snowflake • Tableau • TensorFlow

Figure 9.4 – Vendor and open source tools to implement a Data Fabric solution

Now that we have context on sample tools that could be leveraged in a Data Fabric architecture, let's dive into use cases addressed by a Data Fabric architecture.

Use cases

The business use cases that could be realized using a Data Fabric solution are numerous. They include Customer 360, migration to the cloud, regulatory compliance, data democratization, and more. The theme throughout this book has been the criticality and emphasis on mature Data Governance, where active metadata is the glue that binds distributed data during its life cycle. Data Fabric is a technical approach that is positioned to address a diverse set of use cases.

At the time of writing this book, there has been a big shift toward distributed data management and achieving this via a Data Mesh, an organizational approach with specific data management principles. The accountability and responsibility of data management are moved to each business domain as opposed to a central IT organization leveraged across business domains. It has a federated Data Governance model that is accountable for global data policies and decision-making in order to adhere to government regulations. Data Mesh was introduced and discussed in *Chapter 3, Choosing between Data Fabric and Data Mesh*.

In the next section, we will talk about two use cases – distributed data management via Data Mesh, and regulatory compliance.

Distributed data management and sharing via Data Mesh

There are a lot of discussions today on how to best address the complexities of distributed data management. Large-scale organizations see the value a Data Mesh architecture can offer. The rationale is that it directly targets key pain points by introducing an organizational approach with prescriptive principles. An existing challenge faced by many organizations is the bottlenecks experienced with traditional centralized data management, such as getting access to high-quality governed data. Often, the lack of business domain knowledge and skills within central IT teams managing data lead to a slew of issues, such as untrustworthy data, operational inefficiencies, and lost profit. Data Mesh has taken a stance in addressing these challenges by moving IT management within the business, where the business domain knowledge is in place. While a Data Mesh addresses these challenges, it introduces other challenges, which we will talk about later, in the *Data Fabric with Data Mesh reference architecture* section. The following are the key requirements and personas that need to be addressed and supported in a data architecture.

The key drivers in a Data Mesh approach revolve around four key principles:

- **Domain ownership** – Support a distributed data management model across business domains. Business domains own and manage the life cycle of their domain-based data.

- **Data as a product** – Analytical Data Products are shareable and used by a diverse set of data consumers. Data Products must have the following attributes: discoverable, addressable, understandable, trustworthy, natively accessible, interoperable, valuable, and secure.

- **Self-serve data platform** – A data platform in which business domains manage the life cycle of a Data Product from creation to consumption. The platform provides Self-Service, Data Product development, and sharing, access, processing, and governance of data.

- **Federated computational governance** – An operational governance model that supports a federated organization. It must support interdomain Data Governance requirements such as global policies with local domain execution and enforcement. It must achieve this with a high degree of automation and incorporating *policies as code*.

The personas supported as defined in the Data Mesh are as follows:

- Data Product owners

- Data Product developers

- Data platform product owner

- Data Governance team

- Data Product consumers

- Data platform product owner

- Data platform developer

In the next section, we will talk about a second use case: regulatory compliance.

Regulatory compliance

Regulatory compliance is the need to adhere and conform to a law or regulation. The adherence to these rules by organizations is a mandatory act that, if not completed, leads to huge fines and penalties. Regulations are typically created to protect data in such a way as to avoid its unauthorized use and eliminate the chance of data breach. There are industrial federal, state, and local laws that need to be compiled by organizations. As such, companies are required to show proof of their compliance. This requires data and technology implementations that impact how data is processed, such as its collection, storage, and distribution.

A Data Governance program is at the center of addressing this use case. An example of a regulation is the international regulation of GDPR, which impacts European data. Let's use that as an example to identify requirements and personas that need to be addressed by the data architecture.

The key regulatory compliance drivers are as follows:

- Personal data must be protected (first name, last name, email address, other personal information) during data processing activities for European residents

- Data cannot be exported outside of GDPR-based countries without proper protection

- Data must be protected across its life cycle from creation to end of life (data processing, storage, and access)

- Access to personal data is only granted to those with a justifiable purpose

The data personas (`https://www.gdpreu.org/the-regulation/key-concepts/data-controllers-and-processors/`) supported are as follows:

- Controllers

- Processors

In the next section, we will dive into the required capabilities in a Data Mesh architecture.

Data Mesh multi-plane requirements

A Data Mesh architecture defines three logical architecture layers, referred to as a multi-plane architecture. Each layer has an expected set of capabilities, personas supported, and required interactions. Let's discuss the objectives of this logical architecture.

Multi-plane architecture

A Data Mesh multi-plane logical architecture consists of three planes: mesh experience, Data Product experience, and infrastructure utility. The mesh experience plane oversees and manages, as a whole, the connected Data Products as part of the Data Mesh(es). The Data Product experience plane is granularly focused at the Data Product level and manages the Data Product life cycle. The infrastructure utility plane is the technical infrastructure that supports both Mesh and Data Product experience planes. *Figure 9.5* illustrates a Data Mesh's multi-plane data platform.

Data Mesh's Multiplane Data Platform

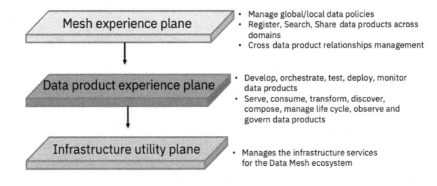

Figure 9.5 – Data Mesh multi-plane data platform

In the next few sections, we will dive into each plane and discuss its objectives. The design of a technical data architecture needs to support the requirements of each plane. One point that I will highlight is that as you review the Data Mesh objectives for each plane, you will find a correlation and commonality to many of the Data Fabric capabilities discussed in the earlier sections that can address these objectives. *You will start realizing the reasoning behind why a Data Fabric and Data Mesh architecture are complementary to one another.*

Let's dive further into each plane.

Mesh experience plane

The mesh experience plane has an end-to-end view of connected Data Products. The capabilities in this plane create, align, and traverse through interdomain Data Product relationships and properties. The capabilities are not isolated to support a specific data domain or Data Product but rather focus on a *big-picture, single-pane* view of the Data Mesh.

The mesh experience plane depends on the Data Product plane. The primary personas supported are the governance team and Data Product owners. The secondary personas are Data Product developers and Data Product consumers.

These are the objectives:

- Manage interdomain/global policies and local policies
- Federate global policies for business domains
- Enforcement of local policies
- Register, search, and share Data Products across domains
- End-to-end Data Lineage
- Cross-data product relationships management

Data Product experience plane

The Data Product experience plane manages the life cycle of a Data Product across the development cycle, maintenance, and sharing, to the end of life. The experiences in this plane are focused on individual Data Products by data domains. It supports Data Product management to share with quality at scale. The capabilities in this plane create a Data Product as an architecture quantum, defined by Zhamak Dehghani as follows:

> *"...smallest unit of architecture that can be independently deployed and managed... It has all the structural components that it requires to do its function: the transformation code, the data, the metadata, the policies that govern the data, and its dependencies to infrastructure."*

The Data Product experience plane depends on the infrastructure utility plane. The primary personas are Data Product owners, Data Product developers, and data consumers.

These are the objectives:

- Develop, orchestrate, test, deploy, and monitor Data Products

- Manage Data Products as code; this includes data pipelines, infrastructure, interfaces, and data policies

- Data Product architecture consists of code, metadata, data, policies, and infrastructure designed as one unit

- A Data Product design must enable the following:

  - Serving data

  - Consuming data

  - Transforming and processing data

  - Discovering, understanding, exploring, and trusting

  - Composing data

  - Managing the life cycle

  - Observing, debugging, and auditing

  - Governing data

Infrastructure utility plane

This plane is focused on the infrastructure that enables the hosting, storage, and compute to process Data Products. The infrastructure utility plane supports both the Data Mesh experience plane and the Data Product experience plane. The primary personas supported are Data Product platform owners and Data Product developers. The secondary personas are Data Product owners, Data Product developers, and data consumers.

These are the objectives:

- Manage the infrastructure services for the Data Mesh ecosystem

- Provide the physical components and resources to enable the Data Mesh and Data Product experience planes

The Data Mesh makes some assumptions in its design. These assumptions need to be reflected in the technical data architecture. Let's discuss what they are in the next section.

Data Mesh assumptions

Assumption 1: Each domain team can have completely different technical platforms and skill sets.

Domain teams have full control and autonomy over the ownership and management of Data Products within their domain. In the Data Mesh's true form to adhere to its principles, it implies that domain teams are free to choose technologies and infrastructure independently for their domain within an organization. In addition, there is an assumption that the technical skill set to fill roles such as the Data Product developers and platform architects would be in place within each domain to fully realize and follow Data Mesh's principles.

I almost think about this in a way where each domain becomes its own mini-company in an organization, with the exception of Data Governance, which requires a federated operational model. This would equate to (in its most complex case) each domain team having completely different technologies, tools, and data storage. Each technical role would have different skill sets and specialties depending on the tools selected. They might also have different funding models that require the separation and management of technical infrastructure components that impact the cost.

Assumption 2: Metadata and policies are embedded into the data product architecture.

A Data Mesh's Data Product architecture requires code, data, metadata, policies, and infrastructure packaged as one unit. This assumes that as part of the sidecar concept in a Data Product, there are a set of capabilities/services that incorporate metadata and policies as part of the code. Metadata and policies are typically managed in separate repositories due to their inherent nature of enabling metadata reusability to understand data and its associated relationships across domains or subdomains. They also have a big-picture view of data (and in the case of Data Mesh, Data Products) across an organization, not just within individual pockets of domains.

In the *Data Product Management* chapter of his book, *Data Management at Scale, Second Edition*, *Piethein Strengholt* highlights the challenges of combining code, data, metadata, and infrastructure, which are summarized as follows. It leads to the creation of a very complex architecture that creates major overhead in its support and maintenance:

- Complex queries would be required across Data Products to have a big-picture view necessary in enabling Data Lineage and Data Quality

- Managing Data Governance such as data privacy, security and ownership that have the same semantics across several physical data representations would create duplication of metadata

- Metadata updates taking place with data, code, and infrastructure would drastically create complexities with keeping metadata updated and in sync

Keeping all these assumptions in mind, a technical architecture can take different forms depending on how strictly you execute Data Mesh principles. You might choose to create variations of the principles, which I call a hybrid Data Mesh approach, that may align better with a business strategy and architecture vision.

In the next section, we will review a reference architecture for a Data Fabric together with a Data Mesh.

Data Fabric with Data Mesh reference architecture

Data Mesh and Data Fabric both aim to achieve similar objectives; their approach is different but complementary. A Data Mesh architecture can build on a Data Fabric's technical approach. There is a misconception in the market today that only one architecture should be selected to execute data management. The reality is that there should be different styles of architecture incorporated into data management to achieve success. As architects, data practitioners, and technologists, we will have different perspectives and points of view that may spark debates. That's part of us learning, sharing, and collaborating with one another and helps us grow as professionals. I offer a point of view of how a Data Fabric architecture can be used together with a Data Mesh architecture.

A Data Fabric's active metadata-driven architecture is a differentiator that complements and accelerates the Data Mesh's principles. *Figure 9.6* represents a reference architecture where Data Fabric and Data Mesh are jointly applied:

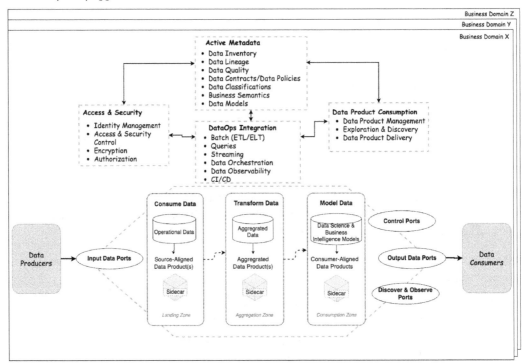

Figure 9.6 – Reference architecture for Data Fabric with Data Mesh

In the next section, we will step through how the architecture works and discuss how it addresses the two use cases we introduced – distributed data management and regulatory compliance.

Reference architecture explained

A Data Product needs to be served, consumed, transformed, discovered, composed, life cycle managed, observed, and governed. Data Product owners, developers, and consumers interact with a Data Product as part of these affordances. Zhamak, in her *Data Mesh* book, defines affordances as properties of a Data Product that have specific interactions between data personas and other Data Mesh components. This is the base of how we will explain the reference architecture.

There are four logical areas that have associated capabilities and functions, which are as follows:

- Access and security
- DataOps integration
- Data Product consumption
- Active metadata

Each area interacts with each other to achieve the required experiences and interactions to support Data Mesh and regulatory compliance personas. Each logical area represents a range of different tools and services necessary in the management of Data Products as part of the Data Mesh ecosystem.

A data marketplace orchestrates the management of a Data Product throughout its life cycle, from development to end of life. It works with other tools, such as the data catalog, knowledge graph, DataOps tools, security tools, and more. It is the starting and end point of a Data Product. Its primary role is to orchestrate the necessary tasks and activities that move a Data Product throughout its life cycle. However, it doesn't always manage them directly but rather enables their orchestration. As an example, the actual development of a Data Product can take place in DataOps-focused tools by a developer and is not necessarily achieved directly in the data marketplace.

A data catalog, together with a knowledge graph, primarily manages the active metadata layer. There are more services/tools that contribute to the generation or collection of metadata, such as DataOps tools. However, both the data catalog and knowledge graph are the key tools accountable for metadata and knowledge management, together with intelligent services that derive the necessary insights.

Event management is another key component. Each time metadata is generated by any of the layers in the architecture, whether data processing, data consumption, or security management, it is collected and sent to the streaming tool. Subscribed consumer tools receive the latest and greatest metadata, such as the data catalog. The data catalog manages business semantics and other metadata at the domain level or the Data Product level. Data models such as metadata models and their configurations to physical data storage locations are captured. A Metadata Knowledge Graph manages all the Data Product relationships, semantic linking, and properties across the entire Data Mesh. It supports the exploration and discovery of Data Products throughout the Data Mesh(es). It leverages active metadata as the navigation vehicle that ties the Data Products together. It also derives new insights based on the knowledge it attains from managing these relationships.

In the next section, we will review all the interactions that take place across each layer in the reference architecture.

Architecture interactions

The DataOps integration layer handles all data processing tasks required in the consumption and serving of Data Products, leveraging technical approaches such as batch, streaming, data replication, distributed querying, and more. Data ingestion activities lead to the creation of Data Mesh's source-aligned Data Products in a landing zone. The DataOps integration layer interacts with the access and security layer to have the necessary authorization and identity management tasks to connect to a source system(s) or provide access to enable the serving of a Data Product. Access and security capabilities interact with the active metadata layer to check and validate data contracts and drive governance. Data contracts contain service-level objectives, data security, data privacy, data retention policies, and allowed purpose of access and use.

The access and security layer enforces the data contract's policies as needed, either dynamically during access or at rest in data storage. In the case of data protection, it encrypts personal data to comply with GDPR. It also grants access to Data Products based on the stipulated policies such as a valid, justifiable business purpose to access personal data. This represents the sidecar policy execution concept of a Data Mesh. The Data Product consumption layer interacts with the active metadata, DataOps integration, and access and security layers to serve Data Products by composing data across different delivery modes (files, distributed/federated queries, events, direct access to tables, APIs, and more) and enable its discovery and understanding. It also captures Data Lineage to address audit requirements necessary in showing regulatory compliance for GDPR. This also includes the management of Data Product access by data consumers for restricted Data Products.

Figure 9.7 demonstrates the four-layered architecture interaction diagram within a Data Fabric and Data Mesh reference architecture:

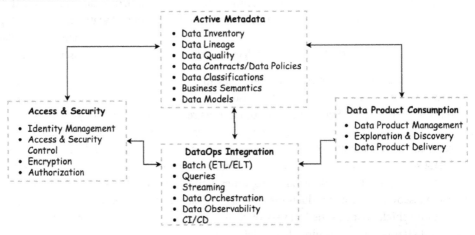

Figure 9.7 – Architecture interaction diagram for Data Fabric with Data Mesh

The Data Mesh's aggregate Data Products focus on applying transformations leveraging interdomain or single-domain source-aligned Data Products to create new Data Products. Transformations can be achieved via a range of approaches, whether programmatic (such as leveraging Python, notebooks, and so on), driven by data as code best practices, ETL/ELT tools, low-code/no-code tools, and more. The active metadata layer is activated each time a Data Product is created, updated, or deleted as part of its life cycle management. The active metadata layer focuses on a shift-left best practice and registers, profiles, and captures Data Quality and lineage for a Data Product. *Data Products are immutable; therefore, a data product is version controlled and keeps track of all data snapshots with associated audit details while keeping the semantics consistent.*

DataOps plays a major role in the development cycle of a Data Product. The DataOps integration layer's capabilities (such as CI/CD, version control, and data observability) support the development life cycle of Data Products. In the case of the Data Mesh's consumer-aligned Data Products, they are created via composability tools that rely on the four layers of the Data Fabric with Data Mesh reference architecture.

Let's discuss the role of a federated operational model.

Federated operational model

In data management, there is a need to support different organizational models. The traditional model has always been a centralized model. Most Data Governance tools and capabilities center around this model. In the case of Data Mesh, it advocates for a federated operational model to manage distributed data. A federated model impacts both Data Governance and infrastructure layout. Put simply, a global team made up of participating business domains and other governance roles is accountable for the Data Governance program.

However, this federated team is not responsible for the execution or enforcement of Data Governance. Each business domain participating in the Data Mesh ecosystem independently owns the enforcement/ adherence to Data Governance policies and manages local policies and local business vocabulary. A technical topology that supports a federated model by enabling the creation and management of global policies and enterprise-wide business vocabulary is needed. The ability to federate or share global artifacts to each business domain in order to enable them to achieve enforcement is necessary. It also should enable the sharing of Data Products across domains as each domain independently manages the life cycle of its Data Products.

Federated approach for Self-Service data platforms

A federated governance team is accountable for managing global policies and standards to which local business domains need to adhere. The assumption here is that each local domain has an independent and dedicated Self-Service platform. In other cases, a hybrid approach might be taken where a small number of domains share a Self-Service platform. This would be considered a hybrid as it wouldn't represent all business domains in an organization, such as in a traditional model, but rather a smaller number of contained business domains. This model might be chosen due to natural organizational alignment, cost factors, and available skill sets. This is not what a Data Mesh advocates, but my point of view here is that there will be variations of these models across organizations as they will represent different progression points, objectives, and maturity levels.

In a true Data Mesh mode, each domain has the autonomy to essentially act as its own mini-company. They manage and own a Self-Serve data platform and Data Product life cycle specific to their domain. However, as Data Products will require consumption representing different domains, it needs to provide a way to share it across a diverse set of consumers representing different domains. Globally managed policies, business vocabulary, and standards created by the federated governance team need to be enforced by each business domain. Having a way to federate policies, standards, and Data Products is a key requirement that must be supported by the capabilities of a Self-Serve data platform.

One way to achieve this is to have capabilities that federate metadata and data from a global Self-Serve data platform to each domain's Self-Serve data platform. *Figure 9.8* illustrates this federated operational model:

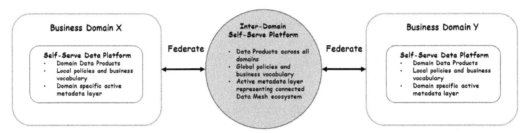

Figure 9.8 – Federated operational model

Alternatively, an organization might choose to have a hybrid operational model that meets in the middle between a federated and centralized Data Governance model. In this model, some of the business domains operate in a true Data Mesh model, while others take a hybrid approach. Such a topology is illustrated in *Figure 9.9*:

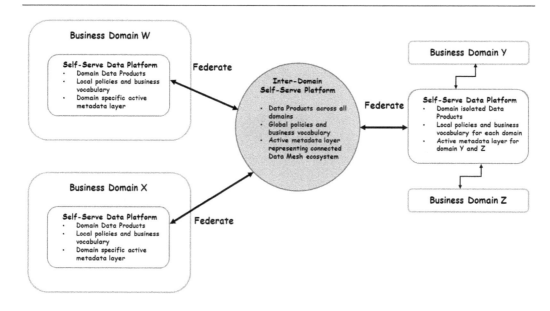

Figure 9.9 – Hybrid operational model

An operational model is driven by the organization's business strategy. Traditional organizational models typically centralize Data Governance. It's important to design a data architecture and have supporting capabilities that are flexible and can support different operational models.

Summary

A technical Data Fabric architecture is modular and composable of several tools and technologies. There are three logical layers in a Data Fabric architecture consisting of Data Governance, Data Integration, and Self-Service.

In this chapter, we reviewed capabilities and the kinds of tools that could be used to implement each layer in a Data Fabric architecture. We have also talked about the requirements and assumptions in two use cases: distributed data management via Data Mesh, and regulatory compliance. A reference architecture was presented that offers a point of view on how to apply both Data Fabric with Data Mesh architectures with a federated operational model. These architectures have similar objectives and call for similar capabilities where Data Fabric is active metadata-driven and takes a more technical approach when compared to Data Mesh, which is more organizationally driven.

In the next and last chapter, we will review industry best practices and recap many of the principles and best practices discussed throughout this book. These best practices represent crucial building blocks in a Data Fabric architecture and general good data architecture.

10
Industry Best Practices

Industry best practices are proven ideas, methods, techniques, and guidelines. In the **Information Technology (IT)** industry, there are many established best practices deemed as successful from lessons learned on what works and doesn't. These experiences have revealed the best course of action on how to achieve data management at a high scale. A best practice achieves quality, greater productivity, lower cost, and increased profit. Gartner defines best practices as "*a group of tasks that optimizes the efficiency (cost and risk) or effectiveness (service level) of the business discipline or process to which it contributes. It must be implementable, replicable, transferable and adaptable across industries.*"

In this chapter, we'll review the top 16 best practices for data management in the IT industry that focus on typical pain points such as data silos, data breaches, and data access bottlenecks. We will revisit many of the best practices applied in a Data Fabric architecture discussed throughout this book. Each best practice is described with a *Why should you care?* explanation of its importance in data management. These best practices represent key knowledge learnings recognized in the industry as success factors with established business value. They can create a path toward a lucrative data monetization strategy.

Top 16 best practices

There are many well-known data management pain points in the IT industry. Some of these pain points are expensive data operations, data silos, bad business decisions from unreliable data, data breaches, and data access bottlenecks. These challenges are faced by enterprises due to the high volume of data and its extensive proliferation. Data is growing at an exponentially high rate each day and it's difficult for organizations to manage data effectively and efficiently while ensuring data is protected and of high quality. Best practices alongside architecture principles and data management frameworks are embedded into a Data Fabric design in order to address many of these pain points. These attributes create a high-performing and mature data architecture capable of handling a diverse set of use cases, across industries, that is optimized to address typical pain points faced by organizations in data management.

The key best practices discussed throughout this book are grouped into four categories:

- **Data Strategy**: A *Data Strategy* represents an organization's plan to become data-driven by reaching a target level of data maturity. Data maturity is the measurement of an organization's expertise in applying data capabilities to achieve business value. Continuous and iterative progress are the objectives, which need to be measured on an ongoing basis.

- **Data Architecture**: A *Data Architecture* is singularly focused on the logical and physical design of an IT data layer within an enterprise architecture. It represents all facets of data management that support an organization's business strategy and capabilities. A Data Fabric design is a type of data architecture.

- **Data Integration and Self-Service**: *Data Integration* unifies data from different systems and is a must-have step in extracting business value from data. Data Integration includes data cleansing, data transformation, and data aggregation and facilitates data sharing within an organization. *Self-Service* empowers data consumers to access and consume trusted business-ready data in a quick and reliable manner without any friction.

- **Data Governance**: *Data Governance* manages data by defining standards, policies, and procedures to achieve data trust. It enables data access and data integrity, and enforces data protection and data security. The goal of Data Governance is to establish business value for an organization with data use.

In the next four sections, we will review key best practices in each category.

Data strategy best practices

Here are key best practices to create and execute a value-driven data strategy. Let's review each one.

Best practice 1

Create a clear and comprehensive data strategy.

A data strategy must align with the business goals and overall framework of how data will be used and managed within an organization. It needs to include standards for how data will be discovered, integrated, accessed, shared, and protected. It needs to address how data will meet regulatory compliance policies, Master Data Management, and data democratization. There needs to be an assurance that both data and metadata have a quality control framework in place to achieve data trust. A data strategy needs to have a clear path on how an organization will accomplish data monetization.

A data strategy is a living document that needs to be continuously updated to align with business goals. It should have a clear maintenance process with frequent reviews and identification of authors and stakeholders that will contribute to the data strategy. This also includes the handling of exceptions to a data strategy process for any one-off decisions in special circumstances. A data strategy document must always be easily assessable, to the point, and understandable.

Why should you care?

A data strategy document is a proactive and thought-leading plan that enables an organization to establish a forward-looking roadmap with an approach to achieve business success. Not having a data strategy document can lead to negative consequences, whether in the form of lost revenue or high operational costs from the lack of readiness and improper planning for the future.

The following are possible consequences of the lack of a data strategy:

- Immature level of organizational readiness
- Incorrect business decisions based on poor-quality or late data
- A tactical and reactive approach to digital transformation, leading to failures from a lack of vision
- Data silos and IT bottlenecks
- Noncompliance with regulations
- Lack of Data Governance, creating risk for data monetization

Let's review best practice 2.

Best practice 2

Establish an organization's data maturity level and progress toward ongoing improvement.

An organization needs to first understand what its current data maturity level is to determine the areas of improvement to create a forward-looking plan. A data maturity assessment offers a position on the current data maturity that serves as an indicator of the health of an organization. A data maturity assessment can be used as a tool to drive continuous improvement by measuring progress. The key thing here is to always strive for continuous improvement to achieve success.

Why should you care?

A data maturity assessment offers a grounded reality of an organization's position in becoming data-driven. Having a sense of where an organization is establishes a starting point and helps set a target to aim for. If you don't know the current data maturity level of an organization, it can lead to the following:

- The misconception of the current level of data maturity, thinking it's higher than it truly is
- Missed improvement opportunities
- No clear view of the current gaps – what's working and not working
- No clear path for how to achieve data monetization

Let's review best practice 3.

Best practice 3

Take a people, process, and technology approach to becoming data-driven.

Data must be recognized as a strategic asset that can evolve into Data Products. A culture change is needed to adopt such visionary thinking and drive change across an organization. It creates better ways of managing data, alignment, and collaboration between business and technical domains. Business processes need to evolve to embrace a mindset that is data-centric. Data management frameworks that focus on quality control, Data Governance, Data Privacy, Data Protection, and Data Security and that achieve operational efficiencies are necessary.

Why should you care?

Technology is a key piece in the puzzle; although, on its own, without applying a people and process strategy, it won't achieve the required business value. You can have an amazing set of technologies that have the potential to create high value with data, although if they're not adopted, they won't yield value. Not having a people and process approach creates tension between business and technical roles. An organization needs a people, process, and technology approach in managing data strategically to avoid failure in becoming data-driven.

Let's review best practice 4.

Best practice 4

Manage data as a strategic asset that evolves into a data product.

The premise here is to stop managing data as a byproduct and create an ecosystem that manages data as a valuable strategic asset that can evolve into a data product. Data producers are accountable for managing the life cycle of data from creation to end of life and ensuring it creates business value along the way for data consumers. This requires data that is governed, trusted, protected, secure, and easily accessible. Move data from technical data assets to Data Products by operationalizing data for high scale sharing.

Why should you care?

Managing data using a value-driven approach enables organizations to more quickly and profitably become data-driven. It addresses the challenges many organizations face on this journey, such as data silos and poor Data Quality. The biggest reason why organizations should implement these best practices is profit! The end goal of managing data with recognized business value in a strategic manner is data monetization. At the end of the day, every organization's main objective, regardless of industry, use case, or size, is to achieve profit.

Let's review the best practices for data architecture.

Data architecture best practices

What are the best practices in designing a quality data architecture? It's a careful balance between positive trade-offs versus negative trade-offs.

The following is a list of best practices in data architecture.

Best practice 5

Assess and document architecture decisions.

A trade-off analysis should be completed when designing a data architecture that carefully evaluates each option, weighing the pros and cons. Documented architecture decisions that capture the options considered and the rationale of going one way or another should be captured as **Architecture Decision Records (ADRs)**. This offers tribal knowledge in the decision-making process, which needs to cover both business and technical reasoning behind the decisions made. ADRs create better collaboration and understanding for future changes required in a data architecture especially as data roles change within teams.

Why should you care?

Trade-off analysis and ADRs avoid the pitfalls typically faced by an architect designing solutions. This includes getting the necessary support to move forward with a data architecture design. Trade-off analysis and ADRs avoid architecture decision anti-patterns that lead to negative results. There are three anti-patterns, coined as follows (source: *Fundamentals of Software Architecture* by *Mark Richards* and *Neal Ford*):

- **Covering Your Assets anti-pattern**: Fear of making the wrong architecture decision, leading to delayed decision-making and impacting the overall development timeline

- **Groundhog Day anti-pattern**: Not knowing the rationale behind an architectural decision, leading to continuous questioning and discussion of the justification for an architecture

- **Email-Driven Architecture anti-pattern**: A lack of implemented architecture decisions caused by ineffective communication

Let's review best practice 6.

Best practice 6

A data architecture needs to be capable of supporting diverse data management styles and operational models.

A data architecture should be designed to support the current data management styles, which as of the time of writing are centralized, federated, and decentralized. The current trend is to adopt a federated and decentralized approach to data management. A data architecture needs to have the robustness and ability to support multiple data management and operational models to provide the necessary business value and agility to support an enterprise's business strategy and capabilities. A traditional centralized operational model can be a starting point within a data architecture; however, it should not be the destination. We are in a world where distributed data management is prominent, and there is a high demand and need for federated and decentralized data management to address the complexities faced in distributed data spread throughout a multitude of systems, tools, and environments.

Why should you care?

Centralized data management has been the best practice for many decades, facilitating and optimizing data management and Data Governance. We are now in a new digital data world where decentralized data management is required. A data architecture needs to be capable of connecting data in an intelligent and cohesive manner and support multiple data management styles in order to address the demand and challenges typically faced by an enterprise. There could be several data management styles and operational models necessary in an enterprise. This is especially important for organizations that have embarked on adopting a decentralized data management model such as Data Mesh. It requires Data Governance to support the inner workings of how data is managed in these operational models. Not having a robust data architecture capable of adapting will create major rework and financial investment, drastically increasing technical debt, not to mention lost productivity and time to value.

Let's review best practice 7.

Best practice 7

Design an architecture that incorporates industry best practices, architecture principles, and quality control frameworks.

Leverage best practices such as these top 16 and others across the industry. Build on the lessons learned from previous architecture approaches and their pitfalls to design a mature and successful design. Adopt data-centric architecture principles that focus on efficient data management, Data Governance, data protection, and security with a focus on quality control. Leverage data management frameworks such as DataOps to achieve productivity and efficiencies in the development and deployment of data across its life cycle to achieve high-scale data sharing and business value.

Here is a recap from *Chapter 6, Designing a Data Fabric Architecture*, on a Data Fabric's architecture principles:

- Data are assets that can evolve into Data Products (TOGAF and Data Mesh): Represents a transition where assets have active product management across its life cycle from creation to end of life, specific value proposition and are enabled for high scale data sharing.

- Data is shared (TOGAF): Empower high-quality data sharing

- Data is accessible (TOGAF): Ease of access to data

- Data Product owner (TOGAF and Data Mesh): The data product owner manages the life cycle of a Data Product, and is accountable for the quality, business value, and success of data

- Common vocabulary and data definitions (TOGAF): Business language and definitions are associated with data

- Data security (TOGAF): Data needs to have the right level of Data Privacy, Data Protection, and Data Security

- Interoperable (TOGAF): Defined data standards that achieve data interoperability

- Always be architecting (Google): Continuously evolve a data architecture to keep up with business and technology changes

- Design for automation (Google): Automate repeatable tasks to accelerate time to value

Why should you care?

There is no value in trying to re-invent the wheel. Learn from the trials and errors of previous approaches. Adopt principles based on the successes of others. Knowing right off the bat what works and what doesn't establishes a higher maturity level. The obvious downfall of not using established principles, best practices, and frameworks is spending non-value-adding time on problems and challenges that have already been figured out before, wasting time on areas already established and resolved instead of spending valuable time building and working on new innovative approaches, and new problems not already figured out.

Let's review best practice 8.

Best practice 8

Create a composable architecture that can easily evolve to include more tools and technologies.

Design an architecture that is robust and composable and can evolve with the latest technologies, market drivers, and business goals. Establish building blocks that serve as the foundation of the architecture. In the case of a Data Fabric, leverage an active metadata and event-driven model to drive automation and quality in data management. Metadata is the glue that ties data together, irrespective of which tools or technologies are in place. Another factor is ensuring the right set of tools and technologies are in place in a composable architecture that aligns with the business goals. Tools and technologies should not be selected based on what the coolest technology is. I've seen this happen plenty of times where technologies are selected based on a "cool factor" with no consideration of what business value they produce.

A data architecture must address the four Vs. The following is a recap from *Chapter 1, Introducing Data Fabric*:

- **Volume**: This focuses on the size of data, which impacts data storage on data systems or devices. Volume has been greatly impacted by the vast amounts of social media and IoT data. Classification of data needs to take place as a first step, and then prioritization. Not all data requires the same level of rigor from a Data Governance perspective.

- **Velocity**: This refers to the speed at which data is collected. Batch processing, real time, and near real time are all factors here. This is critical when it comes to Data Governance as you might not have the complete data context depending on the state of data. Streaming data might not be considered complete until a specific set of events across different times take place. Measuring Data Quality or Data Privacy for streaming data needs an iterative staged approach, which is certainly more challenging than batch processing.

- **Variety**: Data has distinct data formats such as structured, semi-structured, and unstructured data. This dimension impacts the level of implementation in applying Data Governance. As an example, Data Quality analysis cannot be applied directly to unstructured data without some level of classification as a first step.

- **Veracity**: This is where Data Quality comes in. This measures to what degree data can be trusted and relied upon to make business decisions. Of the four Vs, this one is specifically focused on Data Governance while the others impact the approach you take to implement Data Governance.

Why should you care?

Having a robust data architecture that can evolve with the right set of tools and technologies and address the four Vs reduces the cost of development and maintenance and accelerates time to value. It offers a future-proof approach that adapts to changing technologies and other market factors. Not possessing a composable architecture can lead to higher costs in maintenance and ownership, and the risk of not having a competitive advantage.

Let's review best practice 9.

Best practice 9

Leverage advanced technologies such as a Metadata Knowledge Graph, AI, and machine learning to create a sophisticated, active metadata-driven architecture.

A knowledge graph accelerates the value statement in data management. Leverage a Metadata Knowledge Graph to manage business, technical, operational, and social metadata. In *Chapter 7, Designing Data Governance*, I used the human brain as an analogy to explain the power of a Data Fabric's Metadata Knowledge Graph. It can understand and process business context, make associations, and establish new relationships based on existing knowledge. Connect metadata to data and capture the types of relationships across all phases in a data life cycle to reflect an enterprise's connected data ecosystem. Leverage AI and machine learning to automate repetitive Data Governance tasks such as business metadata enrichment and data classification to expedite time-intensive but high-value tasks.

Why should you care?

Eliminate labor-intensive activities with automation. Leverage the Metadata Knowledge Graph, AI, and ML for complex but critical data management activities such as data enrichment, classification, cleansing, and labeling. Data is continuously analyzed to derive insights and make predictions leading to improved performance in data processing and delivery. A Metadata Knowledge Graph, AI, and ML can automate and execute Data Governance functions across Data Quality, Data Lineage, Metadata Management, Master Data Management, Data Privacy, Data Protection, and Data Security.

Next, let's review Data Integration and Self-Service best practices.

Data Integration and Self-Service best practices

Data Integration and Self-Service focus on the development and delivery of data throughout its life cycle, handling inbound and outbound data processing and facilitating data access.

The following is a list of best practices in data processing and Self-Service data access.

Best practice 10

Apply DataOps principles to the development and delivery of data.

DataOps is a best practice framework that accelerates the development of data and quality across its entire life cycle with high efficiency and quality. This is especially important when integrating data across distributed complex systems and environments. Concepts such as quality control, version control, data orchestration, **continuous integration/continuous deployment** (CI/CD), automated testing, automated deployments, and data monitoring are applied to data and metadata. Feedback loops establish an open communication channel for customers to drive ongoing quality improvements.

The following is a list of DataOps Manifesto principles reviewed in *Chapter 4, Introducing DataOps*:

1. Continually satisfy your customer
2. Value working analytics
3. Embrace change
4. It's a team sport
5. Daily interactions
6. Self-organize
7. Reduce heroism
8. Reflect
9. Analytics is code
10. Orchestrate
11. Make it reproducible
12. Disposable environments
13. Simplicity

14. Analytics is manufacturing

15. Quality is paramount

16. Monitor quality and performance

17. Reuse

18. Improve cycle times

Why should you care?

DataOps accelerates operations in building, integrating, and delivering data solutions to drastically cut down the time and cost to deliver data. It expedites data processing and delivery with an emphasis on quality control to quickly achieve business value. Leveraging DataOps principles will enable an organization to achieve a high level of customer satisfaction, an Agile data-driven culture, faster data insights delivery, high Data Quality, fewer system errors, lower cost, high performance and productivity, and efficient collaboration between business and IT.

Let's review best practice 11.

Best practice 11

Embed Data Governance into Data Integration and Self-Service activities to deliver governed trusted data.

Data Governance must be designed and embedded to support the entire life cycle of data across all its phases: create, ingest, integrate, consume, archive, and destroy. Apply the Data Governance pillars—Data Quality, Data Lineage, Master Data Management, Metadata Management, Data Privacy, Protection, and Security—to establish a complete and governed framework. Data Governance needs to be at the center of Data Integration and Self-Service activities to deliver trusted high-quality data that can be acted upon to make business decisions. As discussed in *Chapter 3*, *Choosing between Data Fabric and Data Mesh*, Data Governance requires data that has the following characteristics to derive value: discoverable, addressable, understandable, truthful, natively accessible, interoperable, valuable, and secure.

Why should you care?

Data Governance is a critical success factor in becoming a data-driven organization. Having a Data Governance program saves millions and trillions of dollars from lost revenue and fines by governing data appropriately across its life cycle. Data Governance achieves the following: protects and secures critical data, drives better business outcomes with higher-quality data, executes compliance with regulatory laws, addresses data silos, and enables superior decision-making.

Let's review best practice 12.

Best practice 12

Use data contracts to establish commitment between a data producer and a data consumer.

A data contract includes a service-level and terms of use. This provides a data consumer with clear expectations of the level of data support from a data producer. This includes service-level objectives and agreements that contain aspects such as the frequency of data updates, the handling of data model changes with version control, the turnaround time to resolve defects, and interfaces or delivery modes that will be supported. It also includes data-sharing guidelines a data consumer must adhere to. This has aspects such as what the data consumer is allowed and not allowed to do with the data. It creates commitment from a data producer on what they agree to execute and from the data consumer on what they agree to abide by.

Why should you care?

Data isn't always created with the intention of sharing at a high scale. Data could be created as a one-time activity without any maintenance, resulting in stale data. Data could also be generated without any intention of making high-stake business decisions, leading to untrustworthy data. Data could have been generated in a custom manner for a specific data consumer and does not satisfy multiple data consumers with different use cases. All these factors can lead to extremely negative consequences. Poor decisions are made on low-quality data and overburdened teams are not equipped to handle large-scale data sharing. These are all reasons why it's important to establish clear guidelines between a data producer and a data consumer using a data contract. *You wouldn't buy a house without having a legally binding contract; it's the same case for data.* You need an established level of commitment from both parties to execute as needed with the right set of expectations.

In the next section, let's recap best practices for Data Governance.

Data Governance best practices

Data Governance in today's era enables access to high-quality and trusted data via automation and mature technologies despite the need to enforce security and regulatory requirements. There is consensus across the IT industry on the criticality of Data Governance to be successful in data management.

Let's review best practice 13.

Best practice 13

Define and enforce data policies that address Data Privacy, Data Protection, and Data Security to avoid data breaches.

There should be a high degree of consideration of how an organization addresses Data Privacy, Data Protection, and Data Security. In the event of a data breach, there needs to be a planned course of action to avoid further risk. Data policies need to handle Data Privacy, Data Retention, and Data

Security. Data should be classified as sensitive, government, financial, or other to drive data protection. Automated enforcement policies should be leveraged to take appropriate action in protecting data as needed based on its classifications. Data policies need to be continuously reviewed and kept up to date with the latest laws and regulations.

Why should you care?

Data Privacy, Data Protection, and Data Security policies and enforcement help avoid data breaches and spending millions of dollars on regulatory fines. Creating data policies with a governed enforcement layer across the life cycle of data helps avoid the risk of unauthorized access to data. There are many government-based regulations, such as **General Data Protection Regulation (GDPR)** and the United States' **California Consumer Privacy Act (CCPA)** and **Health Insurance Portability and Accountability Act (HIPAA)**, that must be adhered to. Regulatory compliance is mandatory, otherwise the impact of failing to properly protect and secure data results in millions of dollars in fines, and negative brand impact.

Let's review best practice 14.

Best practice 14

Apply a shift-left active Metadata Management approach.

Metadata needs to be collected at every step in the life cycle of data. Collection should start from the moment data is created all the way to end of life. The active collection of metadata is best exhibited when metadata follows the same path as data throughout its life cycle. Metadata needs to be actively collected, stored, defined, integrated, and standardized alongside data and this should not take place as an after-the-fact activity. There are four types of metadata, which we discussed in *Chapter 7, Designing Data Governance*, that need to be managed:

- **Technical Metadata**: This is the data structure based on where the data is stored. This includes details such as source, schema, file, table, or column names.

- **Business Metadata**: This provides the business context of the data. Typically, this involves capturing terms and definitions of how the business commonly refers to data with descriptions and rules.

- **Operational Metadata**: This is data in a transactional context such as during runtime. An example is data pipeline execution where details such as start and end time, data scope, owner, job status, and so on are captured.

- **Social Metadata**: This type has become more prominent. There is a growing need to put yourself in a customer's shoes to understand what they care about. What are their interests and social patterns? Statistics track the number of times data was accessed and used, by how many users, and for what purpose.

Why should you care?

Metadata Management enables data to be managed at a high scale. Capturing metadata late in the life cycle of data can lead to poor data and lower-quality metadata. The metadata value statement would be diminished if it no longer enabled the discovery, exploration, and understanding of data.

Let's review best practice 15.

Best practice 15

Apply a Data Quality framework to data and metadata to achieve data trust.

Data Quality control is critical in enabling the trust that leads to data monetization. Organizations must trust and rely on data to make strategic business decisions. This requires holistic coverage of data and metadata quality both at rest and in flight. Metadata quality evaluates the integrity of business terms, entity types, attributes, and business rules. Data content quality evaluates the accuracy, completeness, validity, uniqueness, timeliness, and bias of data. Leverage established Data Quality frameworks to execute quality control. One such framework is TIQM from Larry English, discussed in *Chapter 4, Introducing DataOps*. There are several others out there that can be leveraged as well.

Why should you care?

Revenue! Trusted high-quality data reveals trends and enables forecasting analysis leading to increased profit. The right investments and efforts can be put in place to execute and realize potential business opportunities revealed with reliable quality data. Making decisions based on poor-quality data can result in the loss of millions of dollars per year in non-value-adding efforts and lost business opportunities. Imagine approving a high-value mortgage at a low-interest rate to several hundreds of customers thinking they had high credit scores to instead realize they all had poor credit. The impact could be disastrous.

Let's review the last best practice – 16.

Best practice 16

Create a data inventory with business context to democratize your data.

Knowing what data you have, where it is, its business purpose, when it was created and updated, who owns it, and how to access it are key to enabling Self-Service. A data inventory should track raw data, transformed data, and analytical assets such as dashboards, workbooks, reports, ML models, and notebooks. A business vocabulary should provide business context for all data in an inventory. A data inventory should serve two purposes: track enterprise data whether created as a data product to share at a high scale with an established business value proposition and product management or as a data asset or data-related asset to establish its existence and current objectives, which may not be intended for high-scale data sharing.

Why should you care?

Data democratization breaks down data silos by enabling rapid Self-Service access to data. There typically is a long turnaround time from the time data access is requested by a data consumer to the time it takes to be delivered by a data producer. Traditional data access processes contain many manual tasks, require building custom solutions for every data request, lack proper business controls that address data security, and lead to unacceptable turnaround times. These are the typical bottlenecks faced by organizations looking to become data-driven.

Let's summarize what we have covered in this last chapter of the book.

Summary

In this chapter, we have summarized 16 best practices in data management relating to the IT industry. We have recapped best practices across four categories: Data Strategy, Data Architecture, Data Integration and Self-Service, and Data Governance. Each category represents the top best practices that are baked into a Data Fabric architecture. The goals of Data Fabric architecture are to deliver high-quality, governed, protected, and secure data in a high-scale and frictionless manner. *Figure 10.1* depicts a logical Data Fabric architecture covered in *Chapter 8, Designing Data Integration and Self-Service*.

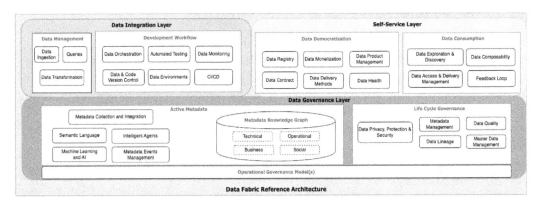

Figure 10.1 – Logical Data Fabric architecture

This concludes the book. You should now have a deeper understanding of and appreciation for Data Fabric architecture. You can comprehend how it can be used together with other architectures and frameworks, the tremendous value statement it offers, its flexible nature, and its ability to support enterprises of all sizes. You understand the major role Data Governance plays in protecting and securing data during its integration and Self-Service. You are now armed with a list of best practices and principles to guide your design and are equipped with the knowledge, direction, and guardrails to design and implement a successful Data Fabric architecture. I wish you great success on your Data Fabric journey.

Index

Symbols

www.packtpub.com

Subscribe to our online digital library for full access to over 7,000 books and videos, as well as industry leading tools to help you plan your personal development and advance your career. For more information, please visit our website.

Why subscribe?

- Spend less time learning and more time coding with practical eBooks and Videos from over 4,000 industry professionals

- Improve your learning with Skill Plans built especially for you

- Get a free eBook or video every month

- Fully searchable for easy access to vital information

- Copy and paste, print, and bookmark content

Did you know that Packt offers eBook versions of every book published, with PDF and ePub files available? You can upgrade to the eBook version at packtpub.com and as a print book customer, you are entitled to a discount on the eBook copy. Get in touch with us at customercare@packtpub.com for more details.

At www.packtpub.com, you can also read a collection of free technical articles, sign up for a range of free newsletters, and receive exclusive discounts and offers on Packt books and eBooks.

Other Books You May Enjoy

If you enjoyed this book, you may be interested in these other books by Packt:

Data Literacy in Practice

Angelika Klidas, Kevin Hanegan

ISBN: 978-1-80324-675-8

- Start your data literacy journey with simple and actionable steps
- Apply the four-pillar model for organizations to transform data into insights
- Discover which skills you need to work confidently with data
- Visualize data and create compelling visual data stories
- Measure, improve, and leverage your data to meet organizational goals
- Master the process of drawing insights, ask critical questions and action your insights
- Discover the right steps to take when you analyze insights

Mastering PostgreSQL 15 - Fifth Edition

Hans-Jürgen Schönig

ISBN: 978-1-80324-834-9

- Make use of the indexing features in PostgreSQL and fine-tune the performance of your queries
- Work with stored procedures and manage backup and recovery
- Get the hang of replication and failover techniques
- Improve the security of your database server and handle encryption effectively
- Troubleshoot your PostgreSQL instance for solutions to common and not-so-common problems
- • Perform database migration from Oracle to PostgreSQL with ease

Packt is searching for authors like you

If you're interested in becoming an author for Packt, please visit authors.packtpub.com and apply today. We have worked with thousands of developers and tech professionals, just like you, to help them share their insight with the global tech community. You can make a general application, apply for a specific hot topic that we are recruiting an author for, or submit your own idea.

Share Your Thoughts

Now you've finished *Principles of Data Fabric*, we'd love to hear your thoughts! Scan the QR code below to go straight to the Amazon review page for this book and share your feedback or leave a review on the site that you purchased it from.

https://packt.link/r/1804615226

Your review is important to us and the tech community and will help us make sure we're delivering excellent quality content.

Download a free PDF copy of this book

Thanks for purchasing this book!

Do you like to read on the go but are unable to carry your print books everywhere?

Is your eBook purchase not compatible with the device of your choice?

Don't worry, now with every Packt book you get a DRM-free PDF version of that book at no cost.

Read anywhere, any place, on any device. Search, copy, and paste code from your favorite technical books directly into your application.

The perks don't stop there, you can get exclusive access to discounts, newsletters, and great free content in your inbox daily

Follow these simple steps to get the benefits:

1. Scan the QR code or visit the link below

https://packt.link/free-ebook/9781804615225

2. Submit your proof of purchase
3. That's it! We'll send your free PDF and other benefits to your email directly

www.ingramcontent.com/pod-product-compliance
Lightning Source LLC
Chambersburg PA
CBHW060133060326
40690CB00018B/3854